GRAPHING CALCULATOR AND EXCEL SPREADSHEET MANUAL

GEX PUBLISHING SERVICES

AF579874

FINITE MATHEMATICS AND CALCULUS WITH APPLICATIONS SERIES

NINTH EDITION

Margaret L. Lial

American River College

Raymond N. Greenwell

Hofstra University

Nathan P. Ritchey

Youngstown State University

PEARSON

Boston Columbus Indianapolis New York San Francisco Upper Saddle River
Amsterdam Cape Town Dubai London Madrid Milan Munich Paris Montreal Toronto
Delhi Mexico City Sao Paulo Sydney Hong Kong Seoul Singapore Taipei Tokyo

This work is protected by United States copyright laws and is provided solely for the use of instructors in teaching their courses and assessing student learning. Dissemination or sale of any part of this work (including on the World Wide Web) will destroy the integrity of the work and is not permitted. The work and materials from it should never be made available to students except by instructors using the accompanying text in their classes. All recipients of this work are expected to abide by these restrictions and to honor the intended pedagogical purposes and the needs of other instructors who rely on these materials.

The author and publisher of this book have used their best efforts in preparing this book. These efforts include the development, research, and testing of the theories and programs to determine their effectiveness. The author and publisher make no warranty of any kind, expressed or implied, with regard to these programs or the documentation contained in this book. The author and publisher shall not be liable in any event for incidental or consequential damages in connection with, or arising out of, the furnishing, performance, or use of these programs.

Reproduced by Pearson from electronic files supplied by the author.

Copyright © 2012, 2008, 2005 Pearson Education, Inc.
Publishing as Pearson, 75 Arlington Street, Boston, MA 02116.

All rights reserved. No part of this publication may be reproduced, stored in a retrieval system, or transmitted, in any form or by any means, electronic, mechanical, photocopying, recording, or otherwise, without the prior written permission of the publisher. Printed in the United States of America.

ISBN-13: 978-0-321-70966-0
ISBN-10: 0-321-70966-7

1 2 3 4 5 6 BRR 16 15 14 13 12

www.pearsonhighered.com

PEARSON

CONTENTS

GRAPHING CALCULATORS

Copyright © 2012 Pearson Education, Inc. Publishing as Addison-Wesley

EXCEL SPREADSHEETS

Copyright © 2012 Pearson Education, Inc. Publishing as Addison-Wesley

Part I

General Instructions for Graphing Calculators

Graphing Calculators

Part I Introduction

Graphing calculators are generally superior to ordinary scientific calculators, not only for graphing (which you would expect), but also for most multi-step calculations. Their screens typically display 7 lines of text, so you can easily see what numbers and functions have been entered. You can easily edit or change these entries, without losing your place in the computation.

Although there are graphing calculators on the market for as little as $40, you would be well advised to use care when buying the less expensive ones, many of which lack useful features that are standard on more expensive models. Before buying a graphing calculator, you should consider which features you are most likely to need. The following chart may be helpful. Many of the features mentioned in the table are discussed in the next part of this supplement.

Feature	Used In FM	Used in CWA	Used in FM and CWA
Table Making Capability	Chapter 1	Chapters 1, 2, 3, 7	Chapter 1
Root Finder (Equation Solver)		Chapter 6	Chapter 14
Financial Functions	Chapter 5		Chapter 5
Matrix Operations	Chapters 2, 4, 10, 11		Chapters 2, 4,
System of Linear Equations Solver	Chapters 2, 10, 11		Chapters 2, 4
Statistical Operations	Chapters 1, 9	Chapters 1, 2, 11, 13	Chapters 1, 9, 10, 18
Maximum/Minimum Finder		Chapters 5, 6	Chapters 13, 14
Intersection Finder	Chapter 1	Chapters 1, 2	Chapters 1, 10
Numerical Derivatives		Chapters 3, 4	Chapters 11, 12
Numerical Integration		Chapter 7	Chapter 15

Current models that have most or all of these features include the TI-83/84 Plus, and the TI-89 graphing calculator. Most graphing calculators have additional features that do not play a role in this book (such as polar coordinates, parametric graphing, and complex numbers). A few, such as the TI-89, can perform symbolic operations such as factoring and finding derivatives.

Copyright © 2012 Pearson Education, Inc. Publishing as Addison-Wesley

Advice on using a graphing calculator

1. **BASICS** Graphing calculators have forty-nine or more keys. Most modern desktop computers have 101 keys on their keyboards. With fewer keys, each key must be used for more actions, so you will find special mode-changing keys such as "**2nd**," "**shift**," "**alpha**," and "**mode**." Become familiar with the capabilities of the machine, the layout of the keyboard, how to adjust the screen contrast, and so on.

2. **EDITING** When keying in expressions, you can pause at any time and use the **arrow keys**, located at the upper right of the keyboard, to move the cursor to any point in the text. You can then make changes by using the "**DEL**" key to delete and the "**INS**" key to insert material. ("INS" is the "second function" of "DEL.") After an expression has been entered or a calculation made, it can still be edited by using the "**ENTRY**" feature. On TI calculators, use 2nd[ENTRY] to return to the previously entered expression. With the new *Math Print* operating system, you can also scroll up to a previous expression and press [ENTER] to paste the expression to the next available line.

3. **SCIENTIFIC NOTATION** Learn how to enter and read data in **scientific notation** form. This form is used when the numbers become too large or too small (too many zeros between the decimal point and the first significant digit) for the machine's display.

4. **FUNCTION GRAPHING**

   A. **Function Memory** Learn how to enter functions in the "**function memory**" (labeled "**Y=**"), and to mark them to be graphed. Depending on the calculator, you can store from 10 to 99 functions in the function memory, so that you don't need to key them in each time you want to use them.

   B. **Viewing Window** Use the "**RANGE**" key (labeled in "**WINDOW**" on some calculators) to determine what part of the coordinate plane will appear on the screen. You must key in the minimum and maximum values for x and y as well as the distance between the tick marks on the axes (for instance, Xscl = 1 and Yscl = 2 means that tick marks will be one unit apart on the x-axis and 2 units apart on the y-axis).

   C. **Graphing Mode** Normally your calculator is set to graph in "**connected mode,**" meaning that it plots the points and connects all of them with a continuous curve (essentially what you usually do when graphing by hand, except that the calculator plots more points). Sometimes (particularly when there should be breaks in the graph, such as a vertical asymptote) connected mode graphing produces misleading or inaccurate graphs. On these occasions, you may want to change the graphing mode to "**DOT,**" so that the calculator will plot the points, but won't connect them.

   D. **Trace** With the "**Trace**" feature, the left/right arrow keys can be used to move the cursor along the last curve plotted, and the values of x and y will be displayed for each point plotted on the screen. On all calculators, if more than one graph was plotted, you can move the cursor vertically between the different graphs by using the up/down arrow keys.

   E. **Zoom** The "**zoom**" feature allows a quick redrawing of your graph using smaller ranges of values for x and y ("**zoom in**") or larger ranges of values ("**zoom out**"). Thus one can easily examine the behavior of a function within the close vicinity of a particular point of the general behavior as seen from far away.

   F. **Plot** With "**Pt-On**" you can move the cursor to any point on the screen and either have the machine plot a single point or have it display the "screen coordinates" of the point. For example, on the TI-83/84 family of graphing calculators, the command "Pt-On(2,3)" will cause this point to be plotted on the graph screen. However, the "Screen coordinates" will usually be approximations of the specified values.

Copyright © 2012 Pearson Education, Inc. Publishing as Addison-Wesley

Solving Equations

1. **EQUATIONS OF THE FORM $f(x) = 0$.** A graphing calculator produces highly accurate approximations of the solutions of equations such as

$$x^3 - 2x^2 + x - 1 = 0.$$

To solve this equation, graph the function $x^3 - 2x^2 + x - 1 = 0$ (see Figure 1). The places where the graph touches the x-axis (the x-intercepts of $y = f(x)$) are the values of x that make $f(x) = 0$, that is, the solutions of the equation. Figure 1 suggests that in this case, there is just one x-intercept, located between $x = 1$ and $x = 2$. On a typical calculator it can be precisely located in several ways. In this text, all figures will show the screen of a TI-83/84 Plus with the classic operating systems, except as otherwise noted.

Graphical Root Finder In the CALC or MATH menu, look for a key labeled "ZERO" or "ROOT." The syntax for this command varies with the calculator. On some calculators, you will be asked to specify a lower and upper bound (that is, numbers on either side of the x-intercept you are seeking) and possibly to make an initial guess. Check your instruction manual for details. The root finder on a TI-83/84 Plus shows that the solution is $x \approx 1.7548777$ (Figure 2).

Zoom-in Use the zoom-in feature of the calculator, or repeatedly change the range settings by hand, so that only a very tiny portion of the graph near the x-intercept is shown. Figure 3, in which the endpoints of the x-axis are 1.7 and 1.8, shows the final window of such a process. The tick marks on the x-axis are 0.01 unit apart and the desired x-intercept is between 1.75 and 1.76, at approximately 1.754 or 1.755. This is very similar to the approximation obtained by the Root Finder.

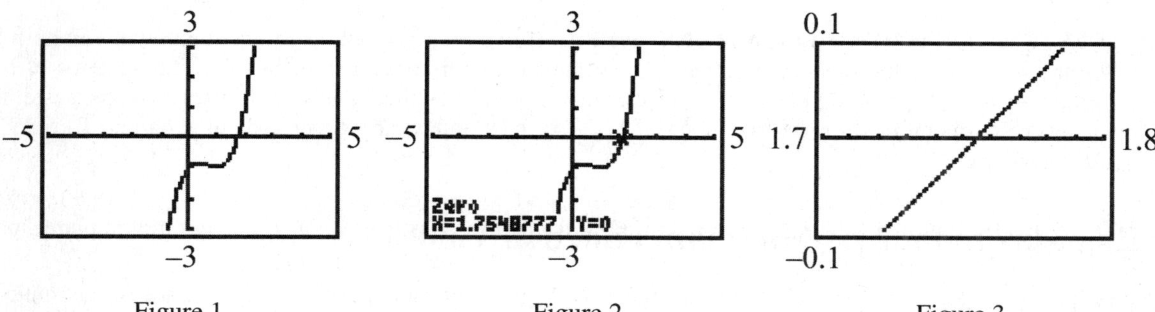

Figure 1. Figure 2. Figure 3.

Algebraic Equation Solver Some calculators have an algebraic equation solver as well, which will solve an equation without the need to graph it first. Check your instruction manual for the correct syntax.

2. **SYSTEMS OF TWO EQUATIONS IN TWO UNKNOWNS** Graphing calculators can solve systems of equations in two variables by finding the intersection points. For example, to solve the system

$$y = x^3 - 2x^2 + x - 3$$
$$y = 4x^2 - 3x - 7,$$

graph both equations on the same screen, as in Figure 4. The points where the graphs intersect are the points (x, y) that satisfy both equations, that is, the solutions of the system. They can be found in two ways.

Copyright © 2012 Pearson Education, Inc. Publishing as Addison-Wesley

Graphical Intersection Finder In the CALC or MATH menu, look for the key labeled "INTERSECTION" or "ISECT." The syntax for this command varies with the calculator. On some calculators, you will be asked to specify the two curves and possibly to make an initial guess. Check your instruction manual for details. The intersection finder on a TI-83/84 Plus (Figure 5) shows that the intersection point to the right of the y-axis has coordinates

$$x = 1.482696 \text{ and } y = -2.654539$$

This is one solution of the system.

Zoom-in Use the zoom-in feature of the calculator, or repeatedly change the range settings by hand, so that only a very tiny portion of the graph near one of the intersections points is shown. Figure 6 (graphed in "grid on" mode) shows the final window in such a process. The grid is determined by the tick marks on the axes, which are 0.01 unit apart. Using the trace feature we estimate that the intersection point has coordinates $(-0.53334, -4.257)$. So the other solution of the system is $x = -0.534$ and $y = -4.257$.

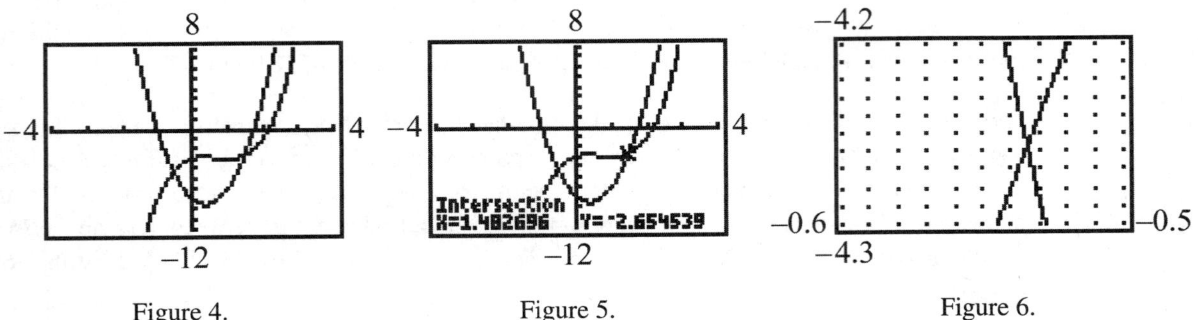

Figure 4. Figure 5. Figure 6.

3. **SYSTEMS OF LINEAR EQUATIONS** Systems of two linear equations in two unknowns can be solved by the methods of the preceding paragraphs. Some calculators have a built-in solver for systems of linear equations and all graphing calculators have matrix capabilities that enable you to solve larger systems by using row operations or matrix inverses. These techniques are described in Chapter 2 of *Finite Mathematics*.

Finding Maximum And Minimum Function Values

The graph of $y = x^3 + x^2 - 3x - 2$ is shown in Figure 7. A graphing calculator can find the local maximum and minimum values of this function (which correspond graphically to the "top of the hill" to the left of the y-axis and the "bottom of the hill" to the right of the y-axis). As with roots and intersections, this can be done in several ways.

Maximum/Minimum Finder To find the local maximum (top of the hill), look in the CALC or MATH menu for a key labeled "MAXIMUM" or "MAX." The syntax for this command varies with the calculator. On some calculators, you will be asked to specify upper and lower bounds and possibly to make an initial guess. Check your instruction manual for details. The maximum finder on a TI-83/84 Plus (Figure 8) shows that the local maximum occurs when $x \approx -1.3874$ and $y \approx 1.4165$.

Copyright © 2012 Pearson Education, Inc. Publishing as Addison-Wesley

Zoom-in Use the zoom-in feature of the calculator, or repeatedly change the range settings by hand, so that only a very tiny portion of the graph near the local minimum point is shown. Figure 9 (in "grid on" mode) shows the final window in such a process. The grid is determined by the tick marks on the axes, which are 0.01 unit apart on the x-axis and 0.001 unit apart on the y-axis. Using the trace feature we estimate that the local minimum point has approximate coordinates (0.721, –3.268). So the local minimum occurs when $x \approx 0.721$ and $f(x) \approx -3.268$.

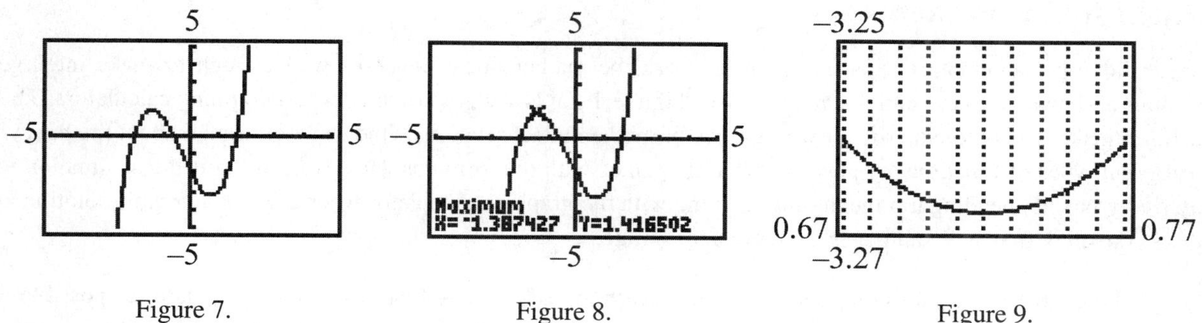

Figure 7. Figure 8. Figure 9.

Programming

Many relatively complicated formulas and processes must be used frequently (for example, the various financial formulas in Chapter 5 and the simplex method in Chapter 7). In such cases you can use the programming feature of a graphing calculator to automate the process. Once a program has been entered into the calculator, it can be called up whenever needed. Typically, the program asks the user to enter the necessary data (for instance, the amount of the loan, the interest rate, and the number of payments) and then quickly produces the desired information (for example, the monthly payment for the loan). Programming syntax and procedures vary from calculator to calculator, so check your instruction manual.

The textbooks website (www.aw.com/LGR) contains useful programs for many graphing calculators. These programs are of two types: programs to give older calculators some of the features that are built-in to newer ones and programs to do specific tasks (such as financial formulas, amortization tables, and the simplex method).

Some Suggestions For Reducing Frustration

We all find ways to make even the simplest machines do the wrong things without even trying. One of the more common problems with graphing calculators is getting a blank screen when a graph was expected. This usually results from not setting the "**Range**" values appropriately before graphing the function, although it could also easily result from incorrectly entering the function.

Another common error is having more right parentheses than left parentheses. To confuse us further, these same calculators think it is perfectly okay to have more left parentheses than right parentheses. For instance, the expression 5(3 – 4(2 + 7) has two left parentheses and one right parenthesis, but it will be evaluated as 5(3 – 4(2 + 7)).

The message "ERR" or "ERROR" appears when a number is too large or when a number is not allowed. If you try to find the 1000th power of ten or divide a number by zero you will most certainly see some kind of error message. When the TI models detect an error, they display a special menu that lists a code number and a name for the type of error. For certain types of errors the choice "**GOTO**" is offered.

Copyright © 2012 Pearson Education, Inc. Publishing as Addison-Wesley

When graphing **rational functions** you sometimes will see strange little "blips" in an otherwise smooth graph. This usually means there is a **vertical asymptote** at that location due to division by zero, but you cannot see the actual behavior there because your "**window**" is too large. "**Zoom in**" on the graph in the region of the irregularity to obtain a better view. Also, as mentioned earlier in this section, one can switch to **dot mode** to eliminate unwanted lines in the graph.

Some Final Comments

While studying mathematics, it is important to learn the mathematical concepts well enough to make intelligent decisions about when to use and when not to use "high tech" aids such as computers and graphing calculators. These machines make it easy to experiment with graphs of mathematical relations. One can learn much about the behavior of different types of functions by playing "what if" games with the formulas. However, in a timed test situation you may find yourself spending too much time working with the graphing calculator when a quick algebraic solution and a rough sketch with pencil and paper are more appropriate.

To get the most return on your investment, learn to use as many features of your calculator as possible. Of course, some of the features may not be of use to you, so feel free to ignore them. A first session of two or three hours with your graphing calculator and your user's manual is essential. Be sure to keep your manual handy, referring to it when needed.

A final word of caution: These calculators are fun to use, but they can be addictive. So set time limits for yourself, or you may find that your graphing calculator has been more of a detriment than a help!

Copyright © 2012 Pearson Education, Inc. Publishing as Addison-Wesley

Part II

Detailed Instructions for Graphing Calculators

Detailed Instructions for Finite Mathematics

This section of Part II contains detailed instructions for using the TI-83 and 83/84 Plus, and TI-89 for *Finite Mathematics*, tenth edition, and *Finite Mathematics and Calculus with Applications,* ninth edition. Chapter 1 is common to both texts, as well as *Calculus with Applications*, tenth edition, and *Calculus with Applications: Brief Version*, tenth edition. Part II is organized by chapters in *Finite Mathematics*; since not all chapters require detailed explanations of graphing calculator use, some chapters are not mentioned here. Instructions are given first for the TI-83 and 83/84 Plus, followed by instructions for the TI-89, where appropriate.

In this manual, section titles from the textbooks are indicated in italics. References are made to specific examples and exercises from the corresponding sections of each chapter, so you should have your textbook nearby as you read through these instructions.

Copyright © 2012 Pearson Education, Inc. Publishing as Addison-Wesley

Chapter 1 Linear Functions

LOCATION IN THE OTHER TEXTS:

Calculus with Applications: Chapter 1
Calculus with Applications, Brief Version: Chapter 1
Finite Mathematics and Calculus with Applications: Chapter 1

Slopes and Equations for Lines.

Entering Lists and Plotting Points.

A set of points can be plotted, as in Example 14, by first entering the coordinates of the points into two lists—one list for the x-values and one for the y-values. On the TI-83 and 83/84 Plus, there are six list names to choose from—L1 through L6. The TI-83 and 83/84 Plus allow more lists to be added with user-defined names, as does the TI-89.

The steps for entering and graphing lists of points are almost identical on the TI-83 and 83/84 Plus. Start by pressing [STAT] [ENTER] to obtain the list editor. Decide which list name will contain the x-values and use the left/right arrow keys to move the cursor to that list. If there is old data in the list that is no longer needed, use the up arrow to move the cursor to the list name and press [CLEAR] [ENTER] to delete the data. Enter the x-values, one at a time, in order, pressing [ENTER] after each. Choose a second list for the y-values and repeat the same process, being careful that corresponding x- and y-values are in the same position within each list (See Figure 1.)

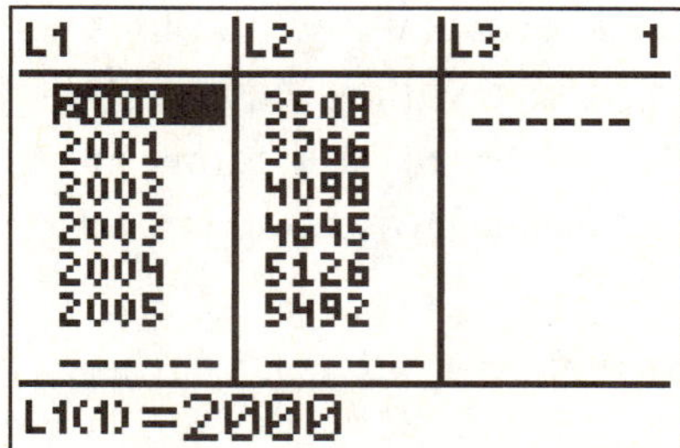

Figure 1.

Copyright © 2012 Pearson Education, Inc. Publishing as Addison-Wesley

To plot the data values, we must go through a few extra steps to make sure that the graph does not contain information we do not want. Begin by pressing [Y=] and "turning off" all functions stored there. To do this, move the cursor to the "=" beside any stored function and press [ENTER] to un-highlight the "=." Now, press [WINDOW] and enter the appropriate range values. To define the plot, press [2ND] [Y=] to obtain the `STAT PLOTS` menu. Press [4] [ENTER] to turn off any other plots, and then return to the `STAT PLOTS` menu. Press [1], [2], or [3] to select a plot name. Turn the plot "on" by pressing [ENTER]. The cursor moves to the `TYPE` line; select the first plot type, the *scatter plot*, by again pressing [ENTER]. Next, the lines titled `Xlist` and `Ylist` need to contain the appropriate list names. On the TI-83 and 83/84 Plus, simply type in the name of the list containing the appropriate values. Finally, select one of three `Marks` that will represent the points on the graph. The plot is now defined; to see it, press [GRAPH]. (See Figure 2, which shows the data plotted in a [2000, 2006] by [3500, 5700] window.)

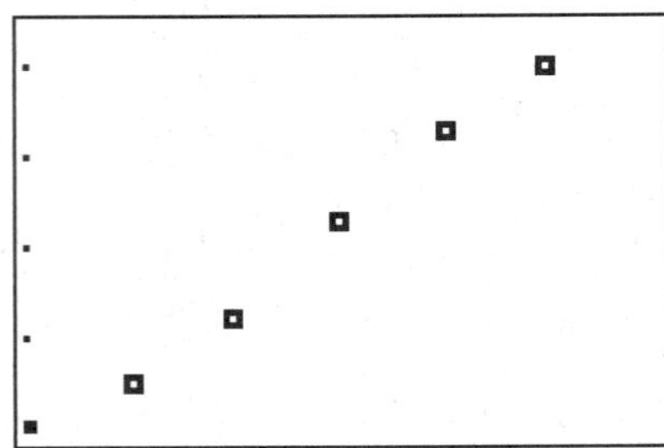

Figure 2.

On the TI-89, press [APPS] and select item 6: `Data/Matrix Editor`, then item 3: `New`. Next to the `Type` option, select `Data`. Type `X` for the `Variable`. Press [ENTER] twice. When the table is displayed, enter the x-values in the first column, pressing [ENTER] after each. Use the right arrow key to move to the second column and enter the corresponding y-values. When finished, press [F2] to enter the `Plot Setup` menu. To define how the plot will be displayed, press [F1], and choose `Scatter` as the `Plot Type`. Press the down arrow key and select one of five marks to represent the points in the graphics screen. Next to `x`, type `C1`, which is the name of the column containing the x-values. Next to `y`, type `C2`. Press [ENTER] twice to accept your changes. Make sure all unnecessary functions are turned off, and enter appropriate window values. Access the `GRAPH` command to see the scatter plot.

Linear Functions and Applications.

Finding Intersection Points Graphically.

In Example 2, of Section 1.2, Figure 11b shows a calculator-generated graph of a pair of supply and demand curves, along with the *equilibrium point*, or *intersection point* of the two graphs. Once the equations of the two functions have been entered into the calculator, all other functions have been "turned off," the appropriate range variables have been chosen, and the graph has been displayed, the calculator can find the intersection point.

Copyright © 2012 Pearson Education, Inc. Publishing as Addison-Wesley

On the TI-83 and 83/84 Plus, while the graph is displayed, press [2ND] [TRACE] to enter the `CALC` menu, then [5] for the `intersect` command. Press [ENTER] twice to select the two curves displayed. Use the left/right arrow keys to move the cursor as close to the intersection point as possible and press [ENTER] again to display the intersection point.

If you are using the TI-89, while the graph is displayed, press [F5] and choose the fifth option, `Intersection`. Press [ENTER] twice to select the two curves displayed. Use the left/right arrow keys to move to a point on the left of the intersection point and press [ENTER]. Move the cursor to a point on the right of the intersection point and press [ENTER] again. The approximate coordinates of the intersection point will be displayed.

Using Solving Equations.

After Example 2 of the text, the authors mention that graphing calculators can be used to solve equations. If an equation contains only one variable, then the `solve` function on the TI-89, can solve it.

Before executing the `Solver` command on the TI-83 and 83/84 Plus, the equation *must* be rewritten so that the left-hand side is "0." Once this bit of algebra is completed, press [MATH] [0] to access the `solver` from the calculator homescreen. Press the up arrow, then type in the right-hand side of the equation. Type in a guess for the solution (see previous paragraph), and press [ALPHA] [ENTER] for the exact solution. (See Figure 3.) Consult the calculator guidebook for an explanation of the other items displayed.

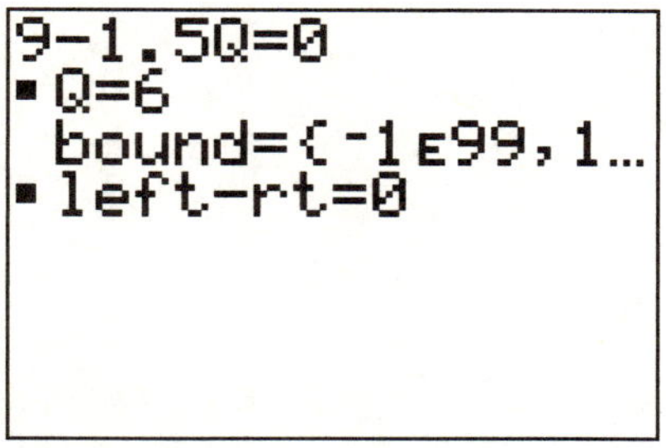

```
9-1.5Q=0
▪Q=6
 bound={-1E99,1…
▪left-rt=0
```

Figure 3.

On the TI-89, from the homescreen, press [2ND] [5] to access the `MATH` menu, and select option `9: Algebra`. Press [ENTER] to access the `solve` command, which is the first item in the `Algebra` menu. Type in the equation to be solved, followed by a comma, then type in the variable for which to solve and close the parentheses. Press [ENTER] to see the solution.

The Least Squares Line

Finding the Least Squares Regression Line.

To complete the example in this section, begin by entering the lists of points as described above. See the instructions beginning just before Figure 17 in Chapter 1 of your text for calculating the regression line on the TI-83 and 83/84 Plus.

Copyright © 2012 Pearson Education, Inc. Publishing as Addison-Wesley

If you are using the TI-89, after entering the lists, press [F5] to access the `Calc` menu. For the calculation type, choose option `5: LinReg`. Press the down arrow and type in the name of the column containing the x-values, then press the down arrow again and type in the name of the column containing the y-values. Press [ENTER] twice to see the coefficients for the least squares line.

Displaying Data and The Regression Line on the Same Graph.

Refer to the discussion above for entering and graphing lists of points. *After* the statistics plot has been defined and the calculator has found the least-squares regression line, the equation can be easily stored for graphing. On the TI-83 and 83/84 Plus, press [Y=] and move the cursor down to an empty function name, keeping the cursor to the right of the "=" sign. Press [VARS] [5] for the `Statistics` variables menu. Press the right arrow twice to move to the `EQ` menu, and select `RegEQ`. The regression equation will be stored in the function menu and its function name will be turned "on." Alternatively, the command "`LinReg L1,L2,Y1`" automatically stores the regression equation under the function name `Y1`. (Other list names and function names may be used here, as necessary.) Press [GRAPH] to see the data and the regression line.

If you are using the TI-89, after entering the lists, press [F5] to access the Calc menu. Set up the window as previously described, but choose a function name in which to store the regression equation, `RegEQ`. After choosing appropriate range values, access the `GRAPH` command to see the regression line and the scatter plot.

Copyright © 2012 Pearson Education, Inc. Publishing as Addison-Wesley

Chapter 2 Systems of Linear Equations and Matrices

LOCATION IN THE OTHER TEXTS:

Finite Mathematics and Calculus with Applications: Chapter 2

Solution of Linear Systems by the Gauss-Jordan Method.

Row Operations.

Once a matrix has been entered, the row operations required for the Gauss-Jordan method can be performed on the calculator. The steps for entering a matrix and performing the row operations are given below.

To enter the matrix of Example 2 on the TI-83, press [MATRX] and press the left arrow key to move to the `EDIT` menu. (On the TI-83/84 Plus, press 2nd [MATRX].) There are ten different matrix names to choose from on the TI-83 and 83/84 Plus. To begin editing matrix [A], press [1]. The dimension of the matrix must be entered first; type in the number of *rows*, press [ENTER], then the number of *columns*, and press [ENTER] again. In Example 2, there are 3 rows and 4 columns. Type in the matrix entries, from left to right, top to bottom, pressing [ENTER] after each one. Press [2nd] [MODE] to return to the homescreen.

On the TI-89, enter the matrix via the `Data/Matrix Editor` from the `APPS` menu. Choose `Matrix` for the Type. Type in a `Variable` name that is not currently in use; this will serve as the name of the matrix being stored. Press [ENTER] twice. Use the down arrow keys to move to the `Row Dimension` and type in the number of rows. Press the down arrow again and type in the `Column Dimension`. Press [ENTER]. Type in the elements of the matrix. Use the arrow keys to move between columns. Return to the home screen.

The row operations for completing the Gauss-Jordan process are located in the [MATRX] `MATH` menu of the TI-83 and 83/84 Plus ([MATRX] followed by the right arrow key), and in the `MATH, Matrix, Row Ops` menu of the TI-89. The individual command names, their locations within the above menus, and their results are summarized below.

Interchange two rows

TI-83: rowSwap	Location:	Option `C`
TI-89: rowSwap	Location:	Option 1

Copyright © 2012 Pearson Education, Inc. Publishing as Addison-Wesley

Multiply a row by a nonzero number

TI-83: *row	Location:	Option E
TI-89: mRow	Location:	Option 3

Add two rows

TI-83: row+	Location:	Option D
TI-89: rowAdd	Location:	Option 2

Add a nonzero multiple of one row to another row

TI-83: *row+	Location:	Option F
TI-89: mRowAdd	Location:	Option 4

Since the syntax for using commands grouped together above is the same for all TI models, we will work through Example 2 using the TI-83 steps.

Once the matrix has been entered, and the calculator has been returned to the homescreen, the matrix name can be copied to an expression or command. On the TI-83, first press [MATRX], followed by the number corresponding to the name of the matrix. (On the TI-83/84 Plus, press [2nd] [MATRX]. On the TI-89, matrix names can be typed in directly from the keypad of the calculator.) To get zeros in rows 2 and 3 of the first column, use the following sequence of commands, pressing [ENTER] after each:

*row+(-3,[A],1,2)→[A]	Multiplies row 1 of [A] by -3; adds the result to row 2.
*row+(-1,[A],1,3)→[A]	Multiplies row 2 of [A] by -1; adds the result to row 3.
*row(1/6,[A],2))→[A]	Multiplies row 2 by 1/6.
row+([A],2,1)→[A]	Adds row 2 to row 1 of [A].
*row(-4,[A],2,3)→[A]	Multiplies row 2 of [A] by -4; adds the result to row 3.
*row(1/6,[A],2))→[A]	Multiplies row 2 by 1/6.

At this point, with the TI-83 and 83/84 Plus, the matrix contains lengthy decimals; care must be taken to avoid round-off errors. Now, or at any time that decimals appear in a calculation, we can convert the matrix entries into fractions by typing [MATRX], pressing the number corresponding to the name of the matrix, then [MATH] [1] [ENTER]. (See Figure 1a). Figure 1b depicts the matrix as shown on the TI-84 Plus with the MathPrint operating system.

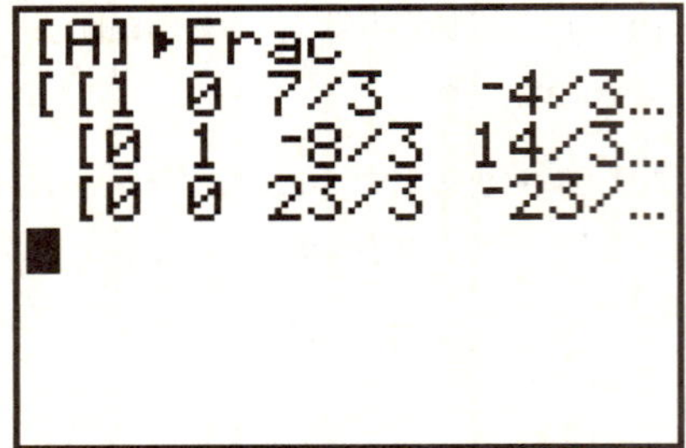

Figure 1a.

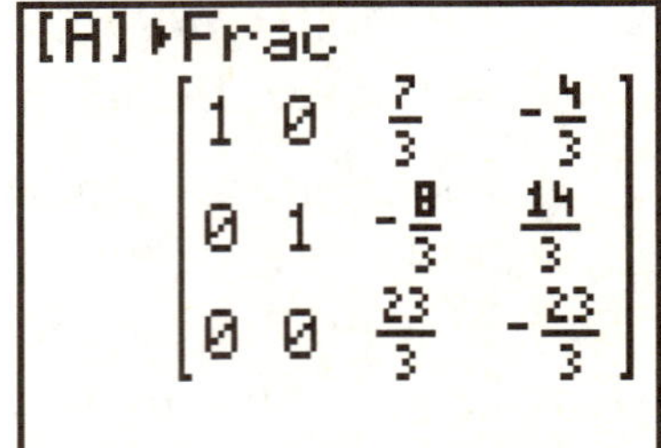

Figure 1b.

Copyright © 2012 Pearson Education, Inc. Publishing as Addison-Wesley

To complete the Gauss-Jordan method, we use the following commands, pressing [ENTER] after each:

*row(3/23,[A],3)→[A]	Multiplies row 3 of [A] by 3/23.
[A] ▸Frac	Converts the matrix entries into fractions.
*row+(-7/3,[A],3,1)→[A]	Multiplies row 3 of [A] by -7/3; adds the result to row 1.
[A] ▸Frac	Converts the matrix entries into fractions.
*row+(8/3,[A],3,2)→[A]	Multiplies row 3 of [A] by 8/3; adds the result to row 2.

The `Frac` command is the first option in the [MATH] menu on the TI-83 and 83/84 Plus. The final result is displayed in Figure 2.

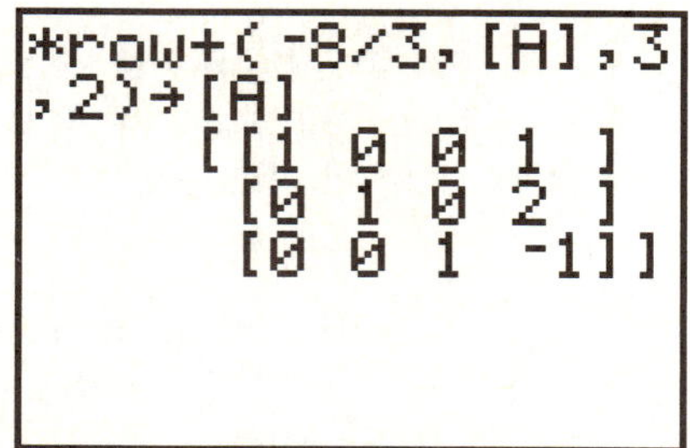

Figure 2.

The `rref` command.

As mentioned after Example 2 of the text, many calculators have a single command which will perform all of the steps of the Gauss-Jordan process at once. First, of course, the matrix must be entered into the calculator as described above.

Once the matrix has been entered into the TI-83 or 83/84 Plus, press [MATRX], press the right arrow key to move to the `MATH` menu, and select option `B`; enter the matrix name from the [MATRX] menu and press [ENTER] to execute the command.

On the TI-89, access the `rref` command in the `MATH MATRIX` menu; type in the name of the matrix, close the parentheses and press [ENTER] to execute the command.

Matrix Inverses.

Calculating the Inverse of a Matrix.

Once a square matrix has been stored, as described previously, the inverse of the matrix, if it exists, can be calculated quickly with any of the TI calculators. From the homescreen, access the name of the matrix on the TI-83 by pressing [MATRX], followed by the number corresponding to the name of the matrix. (On the TI-83/84 Plus, press 2nd [MATRX]. On the TI-89, matrix names can be typed in directly from the keypad of the calculator.) For all models except the TI-89, press [x^{-}] [ENTER]. The `Frac` command can again be used to convert the matrix entries to fractions, when possible.

To find the inverse of a matrix on the TI-89, press [^], and type "`(-1)`". Press [ENTER] to calculate the inverse.

Copyright © 2012 Pearson Education, Inc. Publishing as Addison-Wesley

A Note About Round-off Errors and Matrix Inverses.

Occasionally, when performing calculations involving matrix inverses, one or more entries in the resulting matrix may look like " 1 E -10," or something similar. Keep in mind that this represents 1 times 10 to the -10th power, or 0.0000000001. A result like this in a problem that originally contained no numbers of this form is due to round-off error in the calculator. Usually, it is safe to treat any similar results as "0."

Input-Output Models.

Creating an Identity Matrix.

As mentioned after Example 2 in the text, graphing calculators can be used to determine production for an input-output model. Once matrices A and D have been entered into the calculator, the appropriate identity matrix can be created, and the expression $(I - A)^{-1} D$ can be calculated.

On the TI-83 and 83/84 Plus, an identity matrix can be created using the `identity` command, located in the [MATRX] `MATH` menu. Access this command by pressing [5]. The command must be followed by a number, *n*, which represents the number of rows of the desired identity matrix. Remember to follow this number with [)] In Example 2, the appropriate identity matrix will have dimension 3×3, so we would type "`identity (3)`" for this matrix.

On the TI-89, an identity matrix can be created by accessing the `identity` command, which is the sixth option in the `MATH Matrix` menu. When accessed, you need only enter the number of rows for the matrix, close the parentheses, and continue with the expression to be evaluated.

Copyright © 2012 Pearson Education, Inc. Publishing as Addison-Wesley

Chapter 5 Mathematics of Finance

LOCATION IN THE OTHER TEXTS:

Finite Mathematics and Calculus with Applications: Chapter 5

Simple and Compound Interest.

Compounded Interest on the TI-83 and 83/84 Plus.

The TI-83 is equipped with a TVM Solver, available by pressing [2nd] [x^{-1}] [ENTER]. On the TI-83/84 Plus, press APPS [ENTER]. TVM stands for *time-value-of-money*. This solver will be handy throughout Chapter 5 of your text. There are also several FINANCE commands available, many of which are discussed below. It is very important to remember that these FINANCE commands cannot be used until values have been entered for the variables in the TVM Solver. The variables are as follows:

N Number of payment periods.
I% The annual percentage rate, given as a percent.
PV The present value of the account. If money is being paid *into* the account, PV is entered as a negative number; otherwise, PV is entered as a positive number.
PMT The amount of each payment; if money is being paid *out*, PMT is entered as a negative number; if money is being *earned* or *received*, then PMT is entered as a positive number.
FV Future value of the account.
P/Y Number of payments per year.
C/Y Number of compounding per year.
END/BEGIN If payments are paid at the end of the compounding period, choose END. If payments are made at the beginning of the compounding period, choose BEGIN.

Example 4 of this section of the text can solved by entering the values shown in Figure 1:

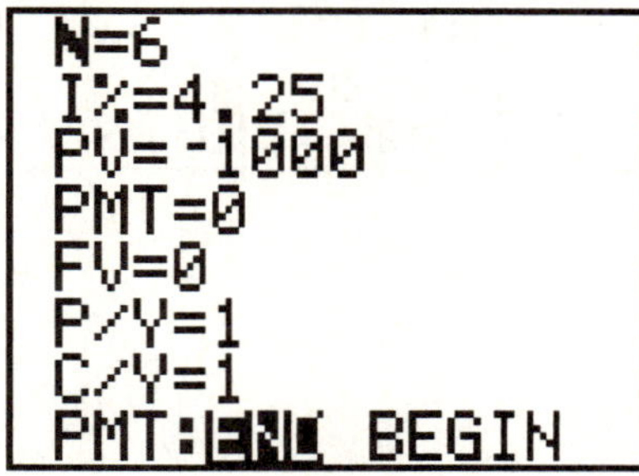

Figure 1.

Copyright © 2012 Pearson Education, Inc. Publishing as Addison-Wesley

In part (a), we are being asked to find the future value of the account, A. To do this with the solver, use the arrow keys to move the cursor beside FV and press [ALPHA] [ENTER] to execute the SOLVE command.

Note: *Often, the calculator's answers will differ slightly from those of the book.*

Effective Interest Rate on the TI-83 and 83/84 Plus.

Other commands in the FINANCE menu can be helpful in this section. The Eff command can be used to find the effective rate in Example 7. From the homescreen, press [2nd] [x^{-1}] and choose option C. (On the TI-83/84 Plus, press APPS and [ENTER] and choose option C. Type in the rate of compounded interest, *as a percent*, followed by a comma, then the number of compounding per year. Press [ENTER] to complete the command. (See Figure 2.)

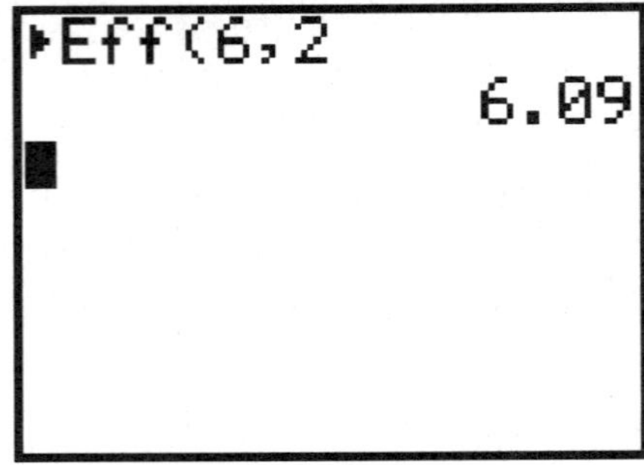

Figure 2.

Present value of an Account on the TI-83 and 83/84 Plus.

Example 9 can also be solved with the TVM Solver. Set **N** = 5, I% = 6.2, PV = 0 (since it is unknown), PMT = 0 (no additional payments will be made into the account), FV = 6000, P/Y = 1, C/Y = 1 (since interest is compounded annually), and choose END. Move the cursor beside PV and press [ALPHA] [ENTER] to find the amount to be deposited. (See Figure 3.)

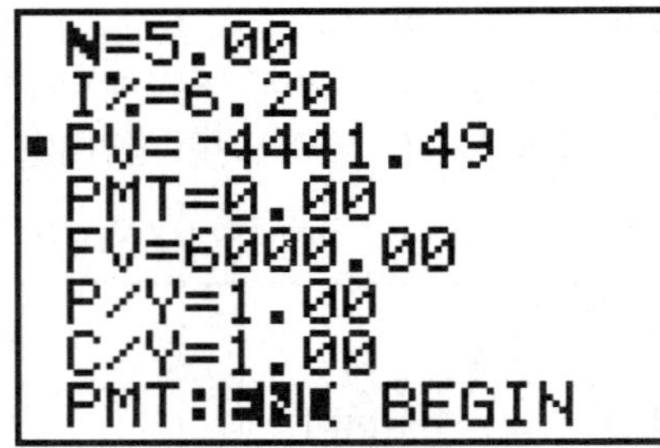

Figure 3.

Copyright © 2012 Pearson Education, Inc. Publishing as Addison-Wesley

Solving for N.

To solve Example 12 of Section 2, set I% = 8, PV = -1, PMT = 0, FV = 2, P/Y = 1, C/Y = 1, and choose END. Move the cursor beside **N** and press [ALPHA] [ENTER] to see that $1 will be worth $2 in approximately 9 years.

There are no built-in commands on the TI-89 model similar to those described in this section, however the functions can be programmed into the calculator. Alternatively, values can be found by entering the appropriate formula in the solve command. For instance, to complete this example on the TI-89, access the solve command in the MATH Algebra menu, type in the equation, a comma, then the variable name, n. Close the parentheses and press ◆ [ENTER]. Also, a FINANCE app for the TI-89 can be downloaded from the TI website.

Future Value of an Annuity.

Ordinary Annuities.

To solve Example 3 of Section 2 we enter the values shown in Figure 4. Solving for FV, the future value of the given sinking fund can be determined.

Figure 4.

To solve Example 4(a) with the TI-83 or 83/84 Plus, set **N** = 12×20 = 240, I% = 7.2, PV = 0 (since it is irrelevant here), PMT = -200, FV = 0 (since it is unknown), P/Y = 12, C/Y = 12 (since interest is compounded monthly), and choose END. Move the cursor beside FV and press [ALPHA] [ENTER] to complete the problem. For part (b) of the same example, replace FV by 130000, and solve for I%. Example 5 can be solved by returning I% to 7.2 and solving for PV.

The solve command on the TI-89, can also be used to solve parts (a) and (b) of Example 4, as previously described. Programs for these models to calculate the future value of an ordinary annuity are included in Part III.

Copyright © 2012 Pearson Education, Inc. Publishing as Addison-Wesley

Annuities Due.

To solve this type of problem, select BEGIN in the last line of the TVM Solver. For instance, Figure 5 represents Example 6 of the text:

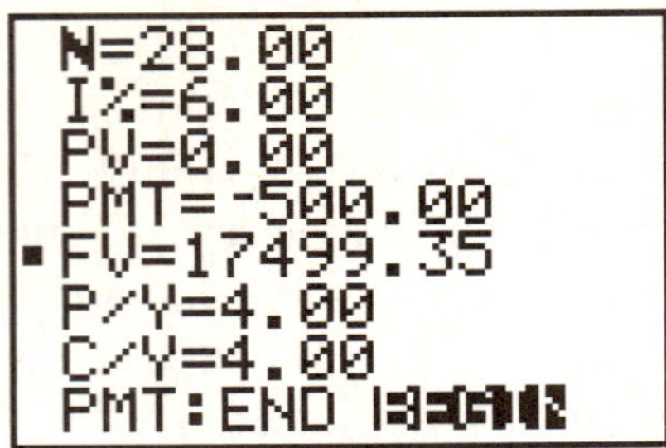

Figure 5.

Again, the TI-89 model does not have built-in functions to perform the financial calculations of this section. However, in Part III of this manual, programs for this model are available for calculating the future value of an annuity. The solve command on the TI-89, can also be used as described previously. This option is also on the FINANCE app for the TI-89 that can be downloaded from the TI website.

Present Value of an Annuity; Amortization.

Present Value of an Annuity on the TI-83 and 83/84 Plus.

The TVM Solver can again be used to perform the calculations necessary in this section of the text. To solve Example 1 of Section 3, press [2nd] [x^{-1}] [ENTER] to obtain the TVM Solver, and enter the values shown in Figure 6. Note that FV is set equal to 0 since it is irrelevant to the problem. Move the cursor beside PV and press [ALPHA] [ENTER] to complete the calculation.

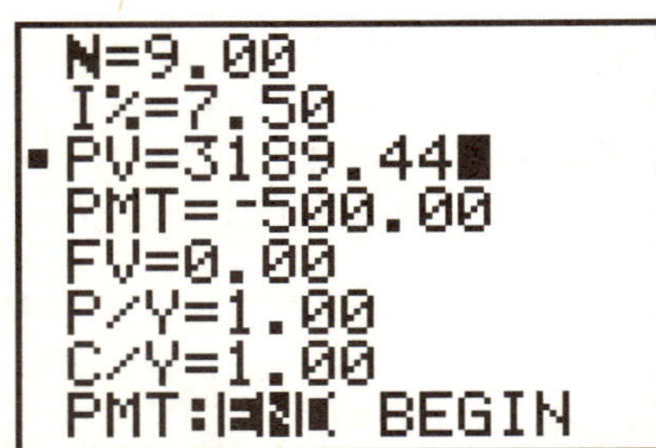

Figure 6.

Amortization.

The TVM Solver and other FINANCE functions of the TI-83 and 83/84 Plus can be used to calculate the size of the periodic payments required to amortize a loan, as well as to find the total amount of interest paid, and to create a partial amortization schedule. The other TI-89 can generate graphical representations of an amortization schedule, and can generate a partial amortization schedule; other functions can be programmed into the calculator. (See Part III.)

Copyright © 2012 Pearson Education, Inc. Publishing as Addison-Wesley

To complete Example 3 (a) with the TI-83 or 83/84 Plus, access the `TVM Solver`. Set **N** = 12×30 = 360, `I%` = 9.6, `PV` = 78000 (the size of the mortgage), `PMT` = 0 (since this is unknown), `FV` = 0 (since it is irrelevant to this problem), `P/Y` = 12, `C/Y` = 12 (since interest is compounded monthly), and choose `END`. Move the cursor beside `PMT` and press [ALPHA] [ENTER] to see that the payment size needs to be $661.56 per month in order to pay off the mortgage in 30 years. For part (b), exit to the homescreen by pressing [2nd] [MODE]. Press [2nd] [x^{-1}] to obtain the `FINANCE` menu, and select option `A`, `Σint`. This function computes the sum of the interest paid between any two payments. Since, in part (b), we want the total interest paid throughout the 30-year mortgage, we need to apply the function to the 1st through 360th payments. So, press [1], then a comma, then [3] [6] [0] and press [ENTER]. (See Figure 7.)

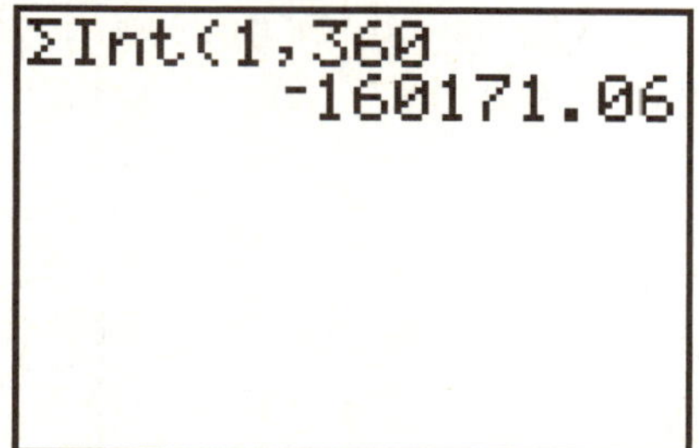

Figure 7.

Remember that the negative sign indicates that the interest is being *paid*, not *earned.* Part (c) can be solved with the same function, this time finding the amount of interest between the 1st and 1st payments; that is, evaluate `Σint(1, 1)`. To find the remaining balance on the loan at any given payment, as described in the discussion following Example 3, we can use the `bal(` function. From the homescreen, press [2nd] [x^{-1}] to obtain the `FINANCE` menu, and select option `9`. Enter the payment number, in this case, 60, and press [ENTER]. The unpaid balance after 5 years, or 60 payments, is $75,122.00.

The simplest way to generate the graph shown as Figure 12(b) of the text will also allow us to generate a partial amortization table for a given loan. To do this, we must change the function setting on the calculator. Press [MODE] and use the arrow keys to move the cursor to the word `Par` in the fourth line; press [ENTER]. The calculator is now set in *parametric mode*. (For more information about this mode setting, see the calculator's guidebook.) Once the proper values have been stored in the `TVM Solver`, press [Y=]. Notice that new function names appear there. Before proceeding, make sure that all `Plots` are "turned off." If a `Plot` is "on", it will be highlighted; turn it off by moving the cursor to the highlighted `Plot` name and pressing [ENTER]; repeat, if necessary, for other `Plots`. Move the cursor back to the first function name, "X_{1T}", and press [X, T, θ,n]; the letter "`T`" should appear. Move to the second function name, "Y_{1T}", and access the `bal` command by pressing [2nd] [x^{-1}] [9]. Press [X, T, θ,n]) to complete the definition. Here, X_{1T} = `T` will represent a payment number and Y_{1T} = `bal(T)` will represent the unpaid balance *after* that payment has been made.

Copyright © 2012 Pearson Education, Inc. Publishing as Addison-Wesley

Press [WINDOW] to set the range values for the graph as Tmin = 0, Tmax = 360, Tstep = 12, Xmin = 0, Xmax = 360, Xscl = 50, Ymin = 0, Ymax = 80000, and Yscl = 10000. (The X and T minimums and maximums should always be the same for this type of graph. The values were chosen to contain the first and last payments. Tstep was chosen as 12 so that each point on the graph will represent the balance at the end of 12 payments, or one year; this speeds up the graphing. The range for the Y values was chosen to include the size of the mortgage.) Press [GRAPH]. You may [TRACE] the graph to see the balance, Y, remaining at the end of each year during the 30-year period, X.

To see the remaining balance at the end of *each* payment, we can use the TABLE function of the calculator. Press [2nd] [WINDOW] to access the TBLSET menu. Set TblStart = 0 and ΔTbl = 1. Make sure that Auto is highlighted for both Indpnt and Depend, and press [2nd] [GRAPH] to see the partial amortization schedule. Scroll through the table with the up/down arrow keys to see the unpaid balance after each payment. (See Figure 8.) Figure 9 is obtained by pressing [GRAPH].

T	X_{1T}	Y_{1T}
1.00	1.00	77962
2.00	2.00	77925
3.00	3.00	77886
4.00	4.00	77848
5.00	5.00	77809
6.00	6.00	77770
7.00	7.00	77731
T=1		

Figure 8.

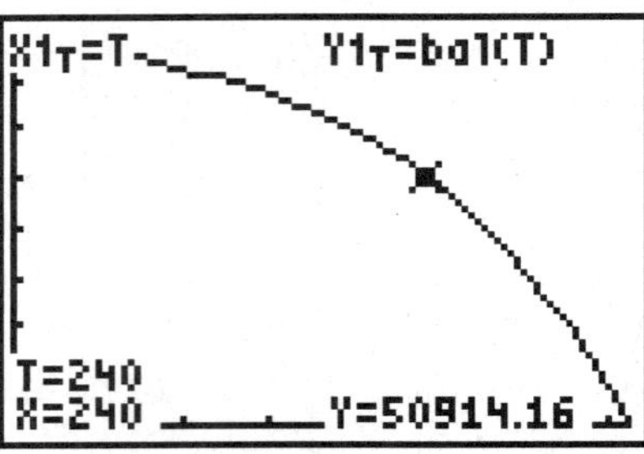

Figure 9.

Similar graphs can be obtained on the TI-89. Put the calculator into *parametric* mode as described above, and set the range variables as indicated. Define xt1 as T, and because this model does not contain the built-in bal command, define Y_{1T} as indicated in the text, $R\left(\frac{1-(1+i)^{-(n-T)}}{i}\right)$; in Example 3, this would be $Y_{1T} = 661.56\left(\frac{1-(1.008)^{-(360-T)}}{.008}\right)$. To view the partial table, access the TblSet menu and store the values indicated previously. Access the TABLE command to view the table.

Copyright © 2012 Pearson Education, Inc. Publishing as Addison-Wesley

Chapter 8 Counting Principles; Further Probability Topics

LOCATION IN THE OTHER TEXTS:

Finite Mathematics and Calculus with Applications: Chapter 8

The Multiplication Principle; Permutations.

Factorial.

Factorial calculations, unless part of another set of commands or a program, should begin from the homescreen. Input the value of n, access the factorial, and press [ENTER].

For example, to calculate 10! on the TI-83 or 83/84 Plus, from the homescreen, type "`10`", then press [MATH], the left arrow key, then [4] [ENTER].

To calculate 10! with the TI-89, from the homescreen, type "`10`", then press [2nd] [5] to access the MATH menu, [7] to access the Probability submenu, then [1] for "`!`". Press [ENTER] to execute the command.

Permutations.

The permutation command on each of the calculators is located in the same area as the factorial command. To calculate a permutation, begin from the homescreen, unless the permutation is part of another function or program. Enter the value of n, access the permutation command, enter the value of r and press [ENTER].

For instance, to calculate P(8, 3) with a TI-83 or 83/84 Plus, as in Example 7 of the text, begin from the homescreen and type "`8`". Press [MATH], the left arrow key, then [2] to access the `nPr` command. Type "`3`" and press [ENTER] to complete the calculation.

To calculate P(8, 3) on the TI-89, from the homescreen press [2nd] [5] to access the `MATH` menu, [7] to access the Probability submenu, then [2] for "`nPr(`". Type "`8`", a comma, "`3`", then `)`. Press [ENTER] to execute the command.

Copyright © 2012 Pearson Education, Inc. Publishing as Addison-Wesley

Combinations.

The combination command is also located in the same area as the two commands above. Execute the combination command in the same way that you execute the permutation command. Example 1, $\binom{8}{3}$, is completed on the TI-83 or 83/84 Plus by typing [8], then [MATH], the left arrow key, then [3] to access the `nCr` command, and [3] [ENTER].

To calculate $\binom{8}{3}$ on the TI-89, from the homescreen press [2nd] [5] to access the `MATH` menu, [7] to access the Probability submenu, then [3] for "`nCr(`". Type "`8`", a comma, "`3`", then `)`. Press [ENTER] to execute the command.

Binomial Probability.

The `binompdf` and `binomcdf` Commands on the TI-83 or 83/84 Plus.

The TI-83 and 83/84 Plus have several probability functions built-in, including the binomial distribution. The appropriate commands are located in the `DISTR` menu of the calculator. The `binompdf` command is option `0`. The command requires you to input the number of trials, n, followed by the probability, p, of one "success," and the number of desired "successes," x. When executed, this command returns the probability of x successes in n trials. The `binomcdf` command requires the same input, but instead returns the probability of x or fewer successes in n trials.

To use these commands to complete Example 4, begin from the homescreen. In part (a), we are asked to find the probability of 1 defective item in a set of 15, where the probability of an item being defective is 0.01. Press [2nd] [VARS] to access the `DISTR` menu, then [0] to access the `binompdf` command. Type "`15`" followed by a comma, "`.01`" followed by another comma, then "`1`" for the number of desired "successes." Press [ENTER] to execute the command. Part (b) of Example 4 asks for the probability of *at most* 1 defective item in a set of 15. The `binomcdf` command will provide this information. It is option `A` in the `DISTR` menu. (See Figure 1.)

```
binompdf(15,.01,
1
          .1303118719
binomcdf(15,.01,
1
          .9903702266
```

Figure 1.

Although the TI-89 does not have a built-in binomial distribution function, it is not difficult to evaluate problems like Example 4(a) with these models, since the binomial probability function, $\binom{n}{x}p^x(1-p)^{n-x}$, can be typed directly onto the homescreen.

Copyright © 2012 Pearson Education, Inc. Publishing as Addison-Wesley

With the TI-89, access the nCr command, then press [1] [5], a comma, then [1]) (. [0] [1] [^] [1]) (. [9] [9] [^] [1] [4]) [ENTER].

The TI-89 can be programmed to complete problems such as Example 4(b). Also, there is a statistical package for the TI-89 that can be downloaded from the TI website that has a binomial distribution function built in.

Copyright © 2012 Pearson Education, Inc. Publishing as Addison-Wesley

Chapter 9 Statistics

LOCATION IN THE OTHER TEXTS:

Finite Mathematics and Calculus with Applications: Chapter 9

Frequency Distributions; Measures of Central Tendency.

Frequency Histograms.

The data in Example 1 of the text can be stored into and displayed by the graphing calculator. In this case, the data should be entered as a single list. (See discussion on page II-3 for assistance with storing lists.) The calculator will use the value input for Xscl as the width of the rectangles in a frequency histogram, so it is important to choose a range of variables that are appropriate for a given data set. In Example 1, Xscl is set equal to 5; thus, each rectangle in the histogram drawn by the calculator will be 5 units in length.

After storing the data list into the TI-83 or 83/84 Plus, we must define a statistics plot. Press [2nd] [Y=] and select a statistics plot name. Turn the plot "on;" beside `Type`, highlight the icon that resembles a histogram. Make sure that the list name under which the data is stored is chosen as the `Xlist`, and that "1" is chosen as the `Freq`. Turn "off" any functions saved in the [Y=] menu and enter the range values, indicated in the text, in the [WINDOW] menu. Press [GRAPH] to see the histogram. The graph can be traced to determine the height and range of each rectangle. (See Figure 1.)

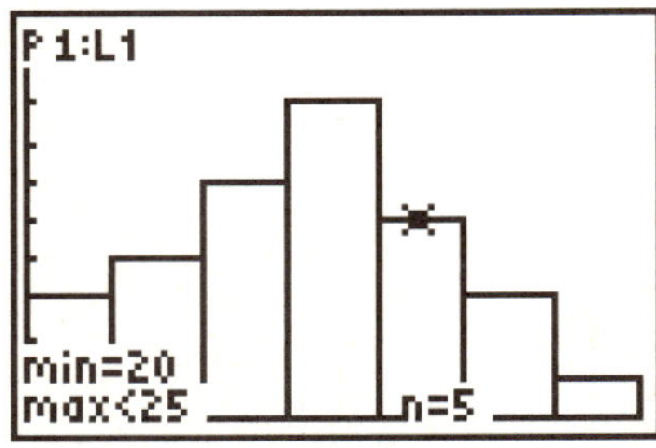

Figure 1.

If you are using the TI-89, set up a data table as described on page II-4 of this manual. Enter the data values in column C1 and, if necessary, the frequencies in column `C2`. Press [F2] to set up the plot, and choose `Histogram` as the `Plot type`. If you have a frequency column, make sure to enter it next to `Freq`. Display the histogram as described on page II-4.

For grouped data, a histogram can be created by storing the *midpoints* of the groups as one list (to be used as the Xlist), and the corresponding frequencies as another (to be used as `Freq`).

Copyright © 2012 Pearson Education, Inc. Publishing as Addison-Wesley

Frequency Polygons.

If data is entered as two lists, one containing the actual values, and the other containing the corresponding frequencies, then a frequency polygon can be drawn on the calculator. On the TI-83 or 83/84 Plus, select the second plot Type and make sure that the appropriate lists are indicated for the data values and the frequencies. Select a Mark to represent the points. Finally, input the appropriate range values and view the graph.

On the TI-89, from the `Data/Matrix Editor`, define the plot and choose xyline as the `Plot Type`, and `C1` for the x-values and `C2` for the y-values. Set the `Freq and Categories` line to No. Otherwise, proceed as described previously.

Calculating Statistics.

Once a data set has been entered as a single list, it is an easy task to find its mean and median on the TI-83 or 83/84 Plus. From the homescreen, access the LIST menu by pressing [2nd] [STAT], use the left/right arrow keys to move to the `MATH` submenu, and select option `3`, `mean`. Type in the name of the list which contains the data and press [ENTER]. The fourth option in the list, `median`, will find the median of a saved list. See Figure 2 for the mean and median of the data set from Example 1 of the text.

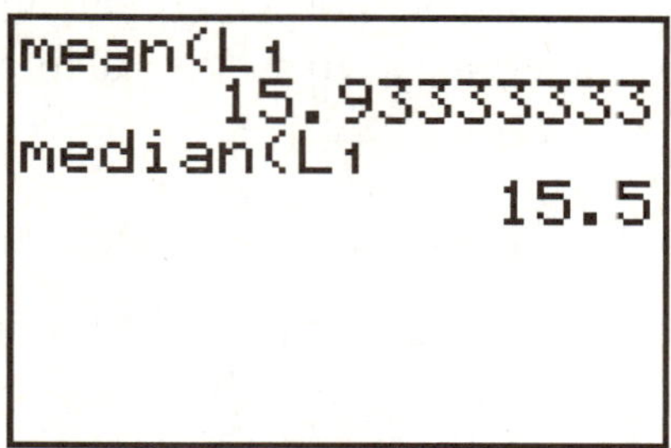

Figure 2.

On the TI-89, from the data screen, press [F5] and choose `OneVar` as the `Calculation Type`. Press the down arrow and enter the column name containing the data values. If there is a list containing frequencies, set `Freq and Categories` to `YES`, and enter the name of the list containing the frequencies next to `Freq`. Press [ENTER] to see the resulting calculations. The first statistic listed, $\bar{x}$, is the mean of the data set; `MedStat` is the median of the data set.

Measures of Variation.

Once a list of data is stored in the calculator, the standard deviation and variance can be calculated quickly. It is very important to remember that the calculator does not know whether the data entered represents a *sample* or a *population*. For that reason, all calculators covered by this manual, *except* the TI-89, will calculate both the population and sample standard deviations. These are distinguished on the TI calculators by σx and `Sx`, respectively.

On the TI-83 or 83/84 Plus, if data is saved in a single list, then option 7, `stdDev`, and option `8`, `variance`, in the LIST menu can be used to calculate sample standard deviation and sample variance of that list.

Copyright © 2012 Pearson Education, Inc. Publishing as Addison-Wesley

If other statistics are required, or if data is entered as two lists (one containing the frequencies), then the `1-Var` Stats command should be used on the TI-83 or 83/84 Plus. From the homescreen, press [STAT] and the right arrow key to move to the CALC submenu; press [1]. Enter the name of the list containing the data. If a frequency list has been stored as well, then type a comma followed by the name of the frequency list. Press [ENTER]. (See Figure 5 of your text.)

Notice that the variance is not included in the list of calculated statistics. Recall that the variance is the square of the standard deviation, so it can be obtained from the information given. The values calculated and shown in Figures 3 and 4 are saved until you either perform a new statistical calculation or reset the calculator. To access these values, press [VARS] [5]. The third and fourth options listed there are the sample and population standard deviations, respectively. To calculate the sample variance from the sample standard deviation, press [3] [x^2] [ENTER].

The methods described previously for calculating the mean on the TI-89 also calculate the sample standard deviations of a data set.

The Normal Distribution.

The TI-83 and 83/84 Plus come equipped with commands to calculate normal probabilities, to find z-scores, and to graph the normal curve. While the TI-89 does not have such built-in functions, it can be programmed to perform the calculations (see Part III); the equation for the normal probability function, $f(x)=\frac{1}{\sigma\sqrt{2\pi}}e^{-(x-\mu)^2/(2\sigma^2)}$, can be graphed by entering it directly into the [Y=] menus of the calculator, with appropriate values substituted for μ and σ.

To calculate the probability $P(a \le z \le b)$ on the TI-83 or 83/84 Plus, use the normalcdf command, which is option `2` in the `DISTR` menu. This command must be followed by the left-hand endpoint, a, a comma, then the right-hand endpoint, b. To calculate $P(-1.02 \le z \le 0.92)$, as in Example 1 (c) of the text, we would enter `normalcdf(-1.02,0.92)` on the homescreen and press [ENTER]. (See Figure 3.)

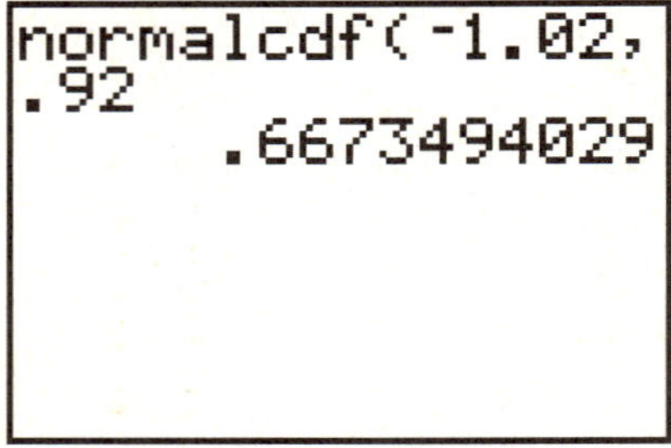

Figure 3.

Copyright © 2012 Pearson Education, Inc. Publishing as Addison-Wesley

To see a graph of the region, as well as to calculate the probability, we can use the `ShadeNorm` command in the `DISTR` DRAW menu. First, we should turn off all unnecessary functions and plots in the [Y=] and `STAT` PLOT menus. Next, we must set an appropriate range for the graph; for the *standard* normal distribution, we might set `Xmin` = -5, `Xmax` = 5, `Xscl` = 1, Ymin = -.1, `Ymax` = .45, and Yscl = .1. You may also want to press [2nd] [PRGM], followed by [ENTER] twice to clear any old graphs or drawings. Now, press [2nd] [VARS] to access the `DISTR` menu and press the right arrow key to move to the `DRAW` submenu. Press [1] to access the `ShadeNorm` command. As with the `normalcdf` command, we must follow this with the left-hand endpoint, a comma, and the right-hand endpoint. Press [ENTER] for the graph and the probability. (See Figure 4.)

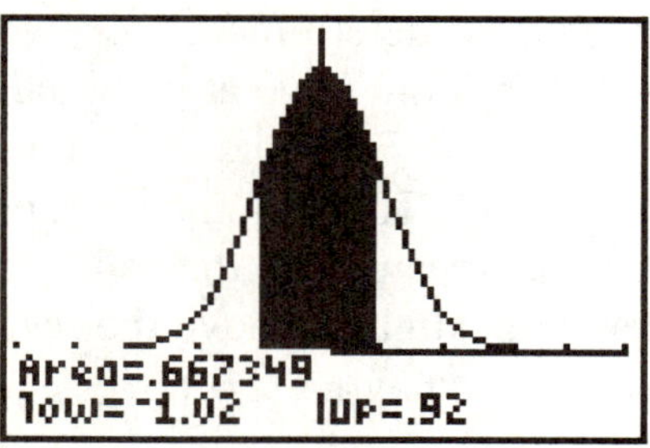

Figure 4.

For problems of the form $P(a \leq z)$ or $P(z \leq b)$, the calculator still requires both a left- and a right-hand endpoint in the `normalcdf` and `ShadeNorm` commands. In Example 1(a), we are asked to find $P(z \leq 1.25)$; since no left-hand endpoint is given, we should choose a very large negative number (a number to the far left of the graph) to use in the calculator command. Your text suggests using -1×10^{99} in this case. Example 1(b) asks us to find $P(1.25 \leq z)$; since no right-end point is given, we should choose a very large positive number to use in the calculator command. In such a case, your text suggests using 1×10^{99}. (See Figures 5 and 6.)

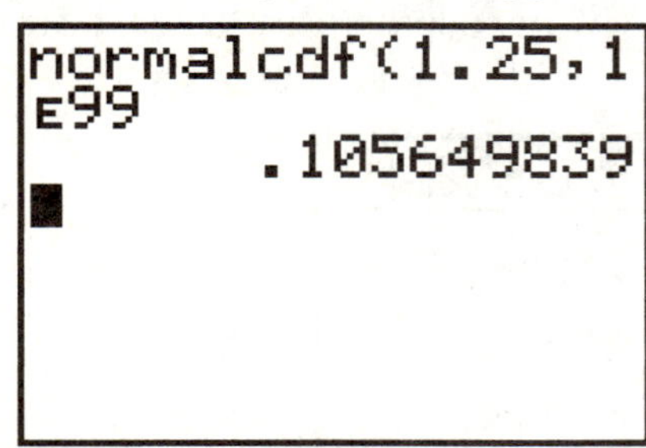

Figure 5.

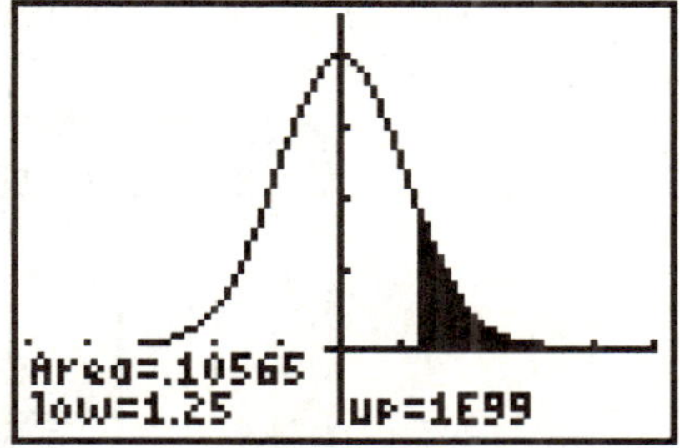

Figure 6.

All three commands discussed here can be used for any normal distribution by adding the mean and standard deviation of the distribution to the commands. For instance, Example 4(a) can be solved by using the command shown in Figure 7 on the following page.

Copyright © 2012 Pearson Education, Inc. Publishing as Addison-Wesley

Figure 7.

In order to use the ShadeNorm command for other normal distributions, we must be careful to select an appropriate range. For `Xmin` and `Xmax`, choose values 4 or 5 standard deviations below and above the mean. To determine `Ymax`, we can use the command normalpdf to find the y-value of the normal distribution curve at its highest point (for a normal distribution, this location is *always* the mean). For the Example 4(b), press [2nd] [VARS] [1] to access this function, and type "`1200,1200,150`". These numbers represent the location of the highest point on the graph, the value of the mean, and the value of the standard deviation for the distribution; press [ENTER]. Use a value slightly above the resulting number as `Ymax`. Ymin can always be chosen as 0, if you do not mind having text cover a portion of the graph.

If you are using TI-89, probabilities involving the normal distribution can be approximated using the `integrate` command in the `MATH Calculus` menu. This command is the second option in the `MATH` Calculus menu. Follow this command with the appropriate formula for the normal distribution, $f(x)=\frac{1}{\sigma\sqrt{2\pi}}e^{-(x-\mu)^2/(2\sigma^2)}$, then a comma, the variable x, another comma, the left-hand x-value, a comma, and finally the right-hand x-value, close the parentheses and press [◆] [ENTER]. Example 1(b), $P(1.25 \le z)$, of the textbook can be approximated as shown in Figure 8. Note that the value 100 was used as the right-hand endpoint, and not 1E99. This is because using too large a number will result in an overflow error on the calculator when it attempts to square the number. Always use a value at least 5σ away from the mean when solving problems similar to Example 1(b) with these calculators.

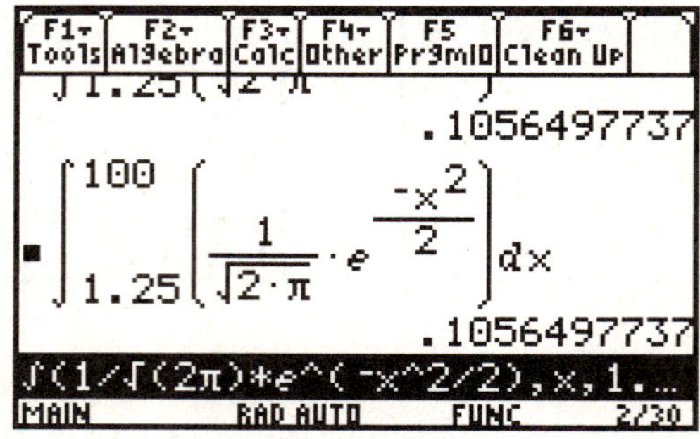

Figure 8.

Also, there is a statistical package for the TI-89 that can be downloaded from the TI website that has a normal distribution function built in.

Copyright © 2012 Pearson Education, Inc. Publishing as Addison-Wesley

Detailed Instructions for Calculus with Applications

This section contains detailed instructions for using the TI-83 or 83/84 Plus and TI-89 with *Calculus with Applications,* tenth edition, *Calculus with Applications: Brief Version*, tenth edition, and *Finite Mathematics and Calculus with Applications*, ninth edition. (Instructions for Chapter 1 of all three texts begin on page II-3.) The section is organized by chapters in *Calculus with Applications*; since not all chapters require detailed explanations of graphing calculator use, some chapters are not mentioned here. Instructions are given first for the TI-83 or 83/84 Plus, followed by instructions for the TI-89, where appropriate.

In this manual, section titles from the textbooks are indicated in italics. References are made to specific examples and exercises from the corresponding sections of each chapter, so you should have your textbook nearby as you read through these instructions.

Copyright © 2012 Pearson Education, Inc. Publishing as Addison-Wesley

Chapter 2 Nonlinear Functions

LOCATION IN THE OTHER TEXTS:

Calculus with Applications, Brief Version: Chapter 2
Finite Mathematics and Calculus with Applications: Chapter 10

Properties of Functions.

Evaluating Functions.

In Example 4(a) of the text, we are asked to calculate $g(3)$, where $g(x) = -x^2 + 4x - 5$. We can check our answers to this and similar problems using the calculator. First, enter the function as Y_1 and return to the homescreen by pressing [2nd] [MODE] to QUIT on the TI-83 and 83/84 Plus, or [HOME] on the TI-89.

To continue with the TI-83 or 83/84 Plus, press [VARS], followed by the right arrow key to move to the Y-VARS submenu. Press [ENTER] to access the FUNCTION submenu, then press the number corresponding to the function name under which $g(x)$ was stored. Type "(3)" and [ENTER] again to evaluate the function. (See Figure 1.)

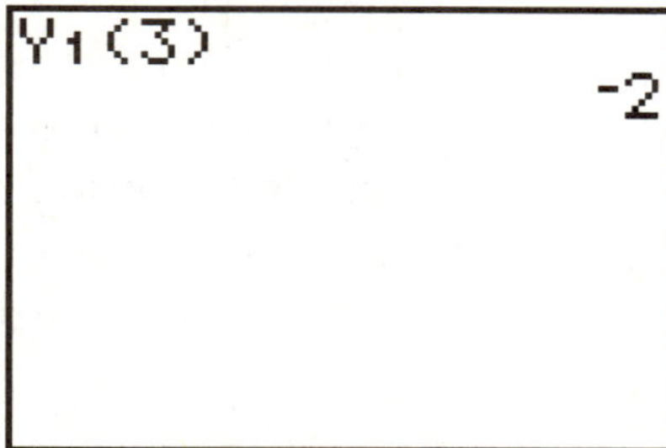

Figure 1.

Using the TI-89, type "y1(3)" and press [ENTER] or ◆ [ENTER] to complete the calculation.

Evaluating Functions from the Graph Window.

While viewing the graph of a function on your calculator, you can evaluate the function at any x-value between Xmin and Xmax, inclusive. If you are using the TI-83 or 83/84 Plus, press [2nd] [TRACE] to access the CALC menu and choose the first command, value. The calculator returns to the graph and prompts for an x-value. To check our answer to Example 4(a) graphically, we would type "3" and press [ENTER]. The coordinates of the point on the graph where $x = 3$ are displayed and the point is marked.

Copyright © 2012 Pearson Education, Inc. Publishing as Addison-Wesley

On the TI-89, while viewing the graph, press [F5] and choose `Value`. Type in the desired x-value and press [ENTER]. The coordinates of the point are displayed, and the cursor moves to that point on the graph.

Similarly, the intersection point of two functions can be found graphically. The instructions for doing this are located in the "Linear Functions" chapter of this manual, on page II-4.

Quadratic Regression.

In Exercise 59 of Section 2 of your text, a quadratic function is fit to a set of data representing Head Start enrollments. The quadratic regression feature of your graphing calculator (`QuadReg` on the TI-83 and 83/84 Plus and TI-89) is in the same location as the linear regression feature. First, data points given need to be entered into two lists—one for the x-values and one for the corresponding y-values. (See page II-13 of this manual for information on entering lists.) Once the data has been entered, return to the homescreen.

On the TI-83 and 83/84 Plus, press [STAT], and the right arrow key to move to the `CALC` submenu. Choose the `QuadReg` command. The command is copied onto the homescreen. Beside it, type the name of the list containing the x-values, a comma, then the name of the list containing the y-values and press [ENTER]. The coefficients for a quadratic function, $y = ax^2 + bx + c$, are calculated and displayed.

On the TI-89, follow the instructions for setting up a statistical calculation within the `Data/Matrix Editor`, as described for linear regression, but choose `QuadReg` as the `Calculation Type`. The values for *a*, *b*, and *c*, as well as for r^2, will be displayed.

If you are using the TI-83 or 83/84 Plus, you can automatically store the regression equation as a function for graphing by making an addition to the end of the command line. For instance, "`QuadReg L1, L2, Y1`" will calculate the quadratic regression equation and store it as `Y1` in the [Y=] menu. The regression equation can then be viewed with the scatter plot of the data. This same addition can be applied to any of the regression commands discussed in this manual.

Graphically Finding the Maximum or Minimum of a Function.

A graphing calculator can be used to determine the maximum of a function, such as the revenue function in part (b) of Example 7. Once the function has been stored, and the graph displayed, we can estimate the maximum (or minimum) for a particular function *within the chosen viewing window*. In the instructions given below, it is *very* important to select appropriate upper/lower bounds for the interval containing the desired maximum; if the interval between the chosen bounds is too wide, an incorrect answer may be obtained. In addition, it is also *very* important to move the cursor as close as possible to the desired maximum when selecting the "guess," when one is required by your calculator, otherwise the calculator may not give you the correct answer.

Copyright © 2012 Pearson Education, Inc. Publishing as Addison-Wesley

While viewing a graph on the TI-83 or 83/84 Plus, press [2nd] [TRACE] for the CALC menu. Select option 4, maximum. You are asked to input a left bound for an interval which includes the maximum value of the function in the viewing window. Use the left arrow key to move the cursor to the left of the maximum and press [ENTER]. You are now asked for a right bound; use the right arrow key to move the cursor to the right of the maximum and press [ENTER]. Finally, you are asked to provide a "guess;" use the left/right arrow keys to move the cursor as close to the desired maximum as possible and press [ENTER]. An estimate for the x-value and the corresponding y-value are displayed.

While viewing the graph of the function on the TI-89, press [F5] and select Maximum. Move the cursor to the left of the point with the maximum y-value and press [ENTER] to select a lower bound, then move the cursor to the right of the point and press [ENTER] to select an upper bound. Press [ENTER] again to see the resulting coordinates of the point with the maximum y-value within the lower and upper bounds.

In order to estimate the minimum point of a function, within a chosen viewing window, follow the steps outlined above, but choose minimum on the TI-83 and 83/84 Plus.

Exponential Functions.

Exponential Regression.

Example 7(b) of this section of the text asks us to find an exponential function that models the given corn production. Another way to find an exponential function that fits a set of data is to use a graphing calculator or computer program with an exponential regression feature. Enter the years as one list in the calculator and the corresponding production levels as another list; return to the homescreen.

On the TI-83 and 83/84 Plus, press [STAT], and the right arrow key to move to the CALC submenu. Choose the ExpReg command (option 0 on the TI-83 and 83/84 Plus). The command is copied onto the homescreen. Beside it, type the name of the list containing the x-values, a comma, then the name of the list containing the y-values and press [ENTER]. The parameters for an exponential function, $y = a \cdot b^x$, are calculated and displayed. (See Figure 2.)

```
ExpReg
 y=a*b^x
 a=1.728250098
 b=1.025404141
 r²=.9914210252
 r=.9957012731
```

Figure 2.

If you are using the TI-89, set up the statistical calculation within the Data/Matrix Editor, as previously described, but choose ExpReg as the Calculation Type. Proceed as with other regression calculations.

Copyright © 2012 Pearson Education, Inc. Publishing as Addison-Wesley

Applications: Growth and Decay; Mathematics of Finance.

Power Regression.

In Exercise 108(d) from the review section of your textbook, you are asked to use the power regression feature of your calculator. This command will calculate the coefficient and exponent for the best-fitting function of the form $y = a \cdot x^b$. Follow the same steps as outlined above for the quadratic and exponential regression commands, this time choosing `PwrReg` on the TI-83 and 83/84 Plus, or `PowerReg` on the TI-89.

Copyright © 2012 Pearson Education, Inc. Publishing as Addison-Wesley

Chapter 3 The Derivative

LOCATION IN THE OTHER TEXTS:

Calculus with Applications, Brief Version: Chapter 3
Finite Mathematics and Calculus with Applications: Chapter 11

Continuity.

Graphing Piecewise Functions.

As indicated in the discussion following Example 3, piecewise functions can be graphed on your graphing calculator. For all TI models *except* the TI-89 (see below), when storing the function under a function name, each piece of the function must be multiplied by a *test* function that indicates the interval corresponding to that piece, and all pieces should be added together. These test functions are located in the TEST menu of each calculator—[2nd] [MATH] on the TI-83 or 83/84 Plus, and [2nd] [5] [8] on the TI-89. Use commands there to type in the piecewise function as it is written in your textbook.

To enter piecewise function on the TI-89, enter each piece as a different function, the use the | key to indicate where this function is used. For instance, y1 = (2x+3) | x< 2 indicates that this function is to be only for x<2.

Definition of the Derivative.

Graphing Tangent Lines.

The tangent line to a function at a given point may be graphed on the same screen as the graph of the function itself, as indicated in the note after Example 1 of your text. Store the function under a function name and turn off other functions and/or plots. Set the range variables so that the x-value of the point where the tangent line is to be drawn is *exactly* halfway between Xmin and Xmax. To duplicate Figure 32 of the text, you might choose Xmin = -4 and Xmax = 2. Then graph the function with your calculator.

If you are using the TI-83 or 83/84 Plus, press [2nd] [PRGM] to access the DRAW menu and choose option 5, Tangent. The graphics screen is displayed again, with the blinking cursor at the center; because of the values chosen above for Xmin and Xmax, the cursor will already be located at the desired x-value. Otherwise, type the x-value of the point after accessing the Tangent command for the tangent line to the function at any point between Xmin and Xmax. Press [ENTER] for the graph of the tangent line. The TI-83 or 83/84 Plus will also display the x-value of the point and the equation for the tangent line at that point. (See Figure 1)

Copyright © 2012 Pearson Education, Inc. Publishing as Addison-Wesley

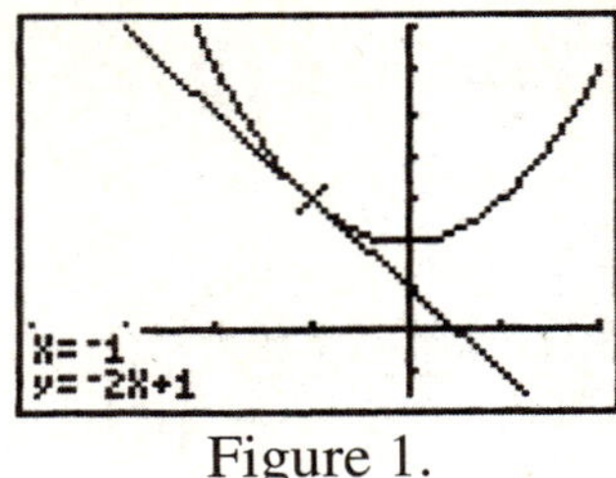

Figure 1.

If you are using the TI-89, while viewing the graph of the function, press [F5] and select option A: Tangent. Type in the x-value of the point whose tangent line you wish to draw.

Numerical Differentiation on the Calculator.

Graphing calculators are equipped with commands that calculate the derivative of a function, when it exists, for a given value of the variable. The TI-83 and 83/84 Plus all calculate derivatives numerically; hence these models only provide approximations of derivatives. The TI-89 calculates derivatives symbolically, much like you would by hand; hence this model provides exact values of derivatives.

If you are using the TI-83 or 83/84 Plus, from the homescreen press [MATH] and choose option 8, nDeriv. Type in the function, followed by a comma, the variable, followed by another comma, and press [ENTER]. Figures 2a and 2b show the result of Example 4(b) of this section of the textbook. (Figure 2b shows this problem on the TI-84 Plus with the MathPrint operating system).

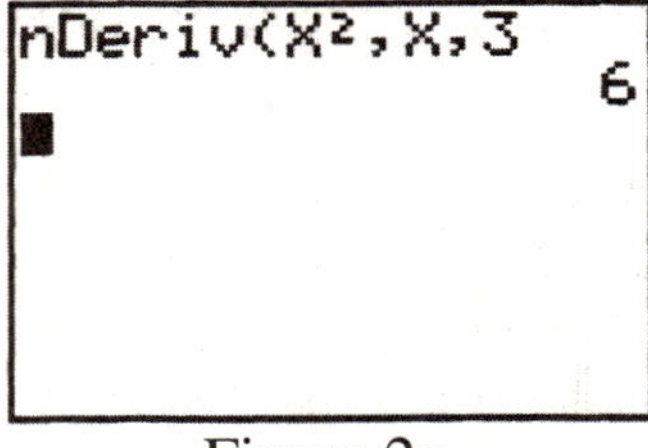

Figure 2a.

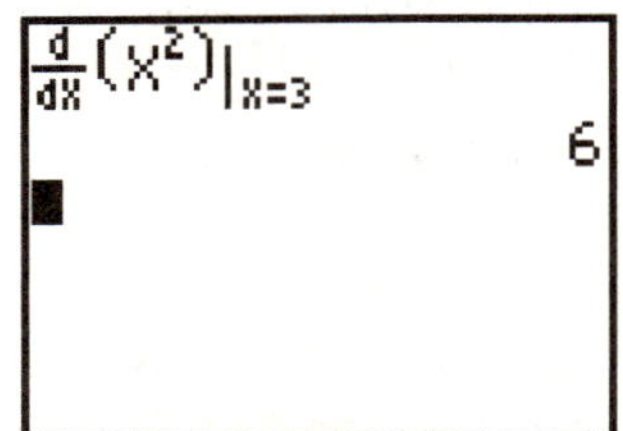

Figure 2b.

Copyright © 2012 Pearson Education, Inc. Publishing as Addison-Wesley

If you are using the TI-89, from the homescreen, press [2nd] [8] to access *d*, the differentiation command. Type in the function (or function name), a comma, then the variable, and close the parentheses. Press [ENTER] to see the derivative of the function. If you wish to evaluate the derivative at a particular x-value, say $x = 3$ then immediately after the differentiation command, you can type [|], then "x=3". Press [ENTER] for the exact value of the derivative.

While viewing the graph of a function, you can also find an approximation for the derivative of the function at a given point. First, store the function, and set the range values as described for graphing tangent lines. On the TI-83 or 83/84 Plus, press [2nd] [TRACE] to access the CALC menu and choose option 6, dy/dx. On the TI-89, press [F5], then select option 6: Derivatives. If you are using the TI-83 or 83/84 Plus or TI-89, you may also type the x-value of the point after accessing the dy/dx command, and press [ENTER] to calculate the derivative of the function at any point between Xmin and Xmax.

Copyright © 2012 Pearson Education, Inc. Publishing as Addison-Wesley

Chapter 6 Applications of the Derivative

LOCATION IN THE OTHER TEXTS:

Calculus with Applications, Brief Version: Chapter 6
Finite Mathematics and Calculus with Applications: Chapter 14

Absolute Extrema.

Checking Answers with the Calculator.

As discussed previously, the maximum and/or minimum of a function, within a certain range of x-values, can be approximated graphically with your calculator. Review the instructions on page II-37, if necessary, and note the importance of choosing appropriate upper and lower bounds for the intervals containing the desired extrema. Also, answers can be checked by using the evaluating function on your calculator, as described on page II-37.

Applications of Extrema.

Approximating Critical Numbers Graphically.

Example 5 of this section involves the equation, $H'(S)=0$, or $2.17\left(\frac{1\ \ln(S+1)}{2\sqrt{S}}+\frac{\sqrt{S}}{S+1}\right)-1=0$, which is quite difficult to solve. If $H'(S)$ is stored under a function name and graphed in a window containing the x-intercept(s), these points, called the *root(s)* or *zero(s)* of the function, can be approximated. If you are using the TI-83 or 83/84 Plus, while observing the graph, press [2nd] [TRACE] to access the `CALC` menu and choose option `2`, which is `zero` on the TI-83 or 83/84 Plus. Use the arrow keys to move the cursor to the left of the desired x-intercept and press [ENTER]. Use the right arrow key to move the cursor to the right of the desired intercept and press [ENTER] again. Finally, move the cursor as close to the intercept as possible and press [ENTER] once more; the approximate x-value of the intercept is displayed. (See Figure 1.)

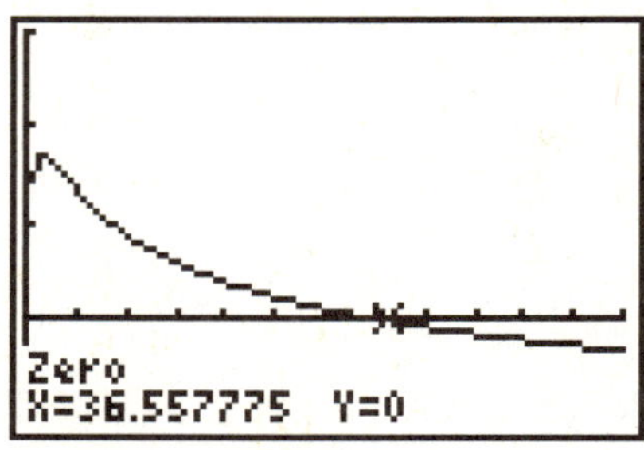

Figure 1.

Copyright © 2012 Pearson Education, Inc. Publishing as Addison-Wesley

On the TI-89, while observing the graph of the function, press [F5] and select the second option, `Zero`. Use the arrow keys to move the cursor to the left of the desired x-intercept and press [ENTER]. Use the right arrow key to move the cursor to the right of the desired intercept and press [ENTER] again.

Alternatively, on the TI-83 and 83/84 Plus and TI-89, the left and right x-values surrounding the desired intercept can be typed in directly, without using the arrow keys, as can the "guess" on the TI-83.

Relative Extrema on the TI-89.

Since the TI-89 is capable of solving equations and differentiating symbolically, it can be used to find the critical numbers of a function. For example, if the function in question is stored as `y1`, then the following command will find the critical numbers of the function: "`solve(d(y1(x),x)=0,x)`".

Alternatively, to find the location of the relative maximum of a function, the command `fMax` in the `MATH Calculus` menu of commands can be used. If the function is stored as `y1`, then the command "`fMax(y1(x),x)`" will find the location of the relative maximum, if one exists. Similarly, the command `fMin` will calculate the x-value of the relative minimum of a function, if a minimum exists.

Copyright © 2012 Pearson Education, Inc. Publishing as Addison-Wesley

Chapter 7 Integration

LOCATION IN THE OTHER TEXTS:

Calculus with Applications, Brief Version: Chapter 7
Finite Mathematics and Calculus with Applications: Chapter 15

Area and Definite Integrals.

Summation.

The discussion following Example 2 demonstrates how lists can be used with the TI-83 and 83/84 Plus to gather the information necessary to approximate a definite integral and to calculate the approximation. To store the headings for list `L2`, after `L1` has been stored, use the arrow keys to move the cursor to the top of the list, so that the list name, `L2`, is highlighted. Type the desired expression for the list (in this case, "`-.5+L1`") and press [ENTER]; the corresponding values for L2 are calculated and displayed. Repeat for `L3`.

On the TI-89, these calculations can be done directly, using the Σ command in the `MATH Calculus` menu. From the homescreen, access this command, then type in the function (or name of the function), followed by [|], then "`(x=-.5+n)`". Multiply this by the size of Δx, then type a comma. Type the variable for the index, `n`, another comma, the first value of `n`, another comma, and the last value of `n`. Close the parentheses and press [ENTER]. For Example 2, the command should appear as follows: "`Σ(2x|x=(-.5+n)*1,n,1,4)`".

The Area Between Two Curves.

Approximating Definite Integrals.

Option `9` in the [MATH] menu of the TI-83 and 83/84 Plus, `fnInt`, can be used to approximate the value of a definite integral. The command must be followed by the function to be integrated (or the function name under which it is stored), a comma, the variable, followed by another comma, the lower limit of the integral, a comma, then the upper limit of the integral. Pressing [ENTER] executes the command. For instance, the integral in Example 4 (b) can be approximated by using the command shown in Figure 1a on the next page. If using the TI-84 Plus with the MathPrint operating system, the definite integral function is accessed in the same manner, and then the equation is filled in as shown on the following page in Figure 1b.

Copyright © 2012 Pearson Education, Inc. Publishing as Addison-Wesley

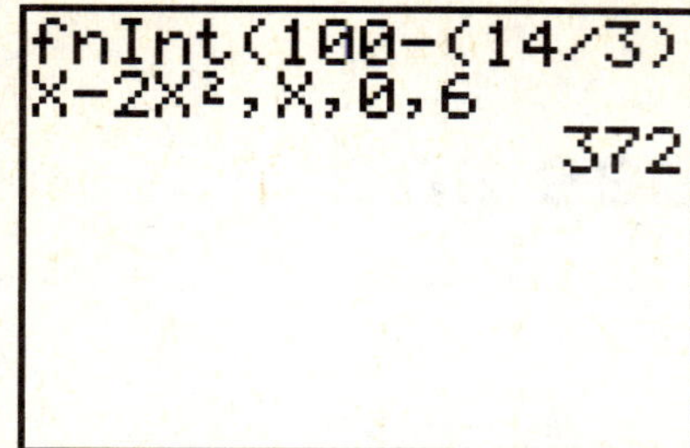

Figure 1a.

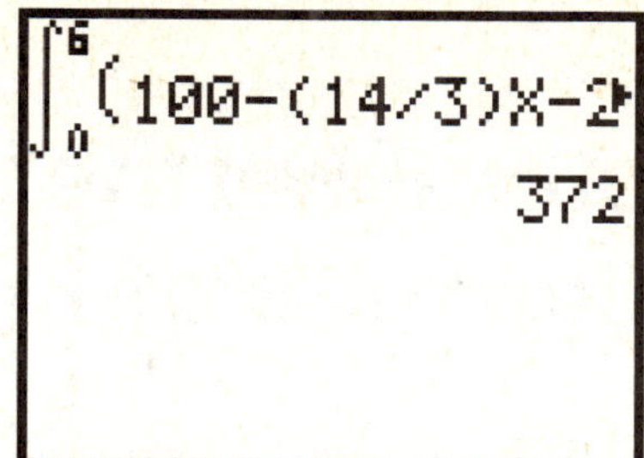

Figure 1b.

If you are using the TI-89, the integral command, accessed by typing [2nd] [7], can be used to find exact values of definite integrals, as well as to evaluate indefinite integrals. To evaluate $\int 100-\frac{14}{3}x-2x^2\,dx$, access the integral command, type in the function (or name of the function), a comma, then the variable name. Press [ENTER] to see the indefinite integral. To evaluate $\int_0^6 100-\frac{14}{3}x-2x^2\,dx$, type in the following command: "`∫(100-(14/3)x-2x^2,x,0,6)`". Press [ENTER] to see the exact value of this definite integral, or [◆] [ENTER] for a decimal approximation.

Copyright © 2012 Pearson Education, Inc. Publishing as Addison-Wesley

Chapter 8 Further Techniques and Applications of Integration

LOCATION IN THE OTHER TEXTS:

Calculus with Applications, Brief Version: Chapter 8
Finite Mathematics and Calculus with Applications: Chapter 16

Integration by Parts.

Approximating Definite Integrals Graphically.

The ∫f(x)dx command of your calculator will approximate a definite integral of a graphed function, where the upper and lower limits of integration are between Xmin and Xmax. On the TI-83 and 83/84 Plus, and TI-89, this command also shades the region corresponding to the definite integral.

To use this command on the TI-83 or 83/84 Plus or TI-89, first store the function and set appropriate range variables. While viewing the graph of the function, access the ∫f(x)dx command; it is option 7 in the CALC menu on the TI-83 and 83/84 Plus, and it is found on the TI-89 by pressing [F5] and selecting option 7. Type in the lower limit of integration, press [ENTER], then type in the upper limit and press [ENTER] again. The definite integral is approximated and the corresponding region shaded. Figure 1 below represents the results of this process, using the integral from Example 4 of this section of your text.

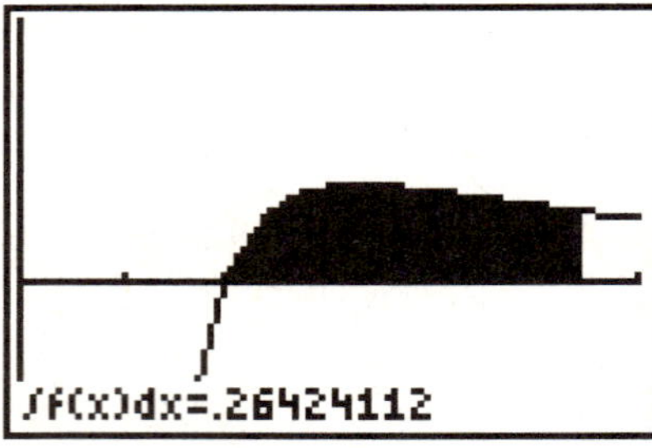

Figure 1.

Improper Integrals

Improper Integration on the TI-89

Improper integrals are set up on the TI-89 in the same manner as a definite integral. Use the ∞ symbol to enter plus or minus infinity, where necessary.

Copyright © 2012 Pearson Education, Inc. Publishing as Addison-Wesley

Chapter 10 Differential Equations

Solutions of Elementary and Separable Differential Equations.

Logistic Regression on the TI-83 and 83/84 Plus and TI-89.

Exercises 40, 47, 48, and 52 of this section of your textbook require the use of the logistic regression command of the TI-83/84 Plus and TI-89. To solve these four exercises, enter the x-values in one list and the y-values in another. (If necessary, review page II-13 of this manual for instructions on entering lists.) Define a statistics plot to graph the two lists, turn off other functions, set appropriate range values, and press [GRAPH].

To determine the logistic function with the TI-83 and 83/84 Plus, return to the homescreen, press [STAT], and move to the CALC submenu by pressing the right arrow key. The Logistic regression command is option B. After choosing this command, type in the name of the x-list, a comma, and the name of the y-list; press [ENTER]. Because the mathematics behind the logistic regression command are complicated, the calculator takes a few seconds to complete the calculations.

To determine the logistic function with the TI-89, from the Data/Matrix Editor, press [F5] and select Logistic as the Calculation Type. Proceed as described for other regression calculations on the TI-89.

Euler's Method.

Calculating y_n.

The graphing calculator method described in Example 1 of this section applies to all TI models covered by this manual. Begin by rewriting the differential equation so that the y' term is isolated on the left-hand side. Then, enter the right-hand side of the equation as Y_1 in the [Y=] menu. Return to the homescreen and store the initial values of X and Y, respectively. On a new line, type "X+", the value of h, press the [S TO▷] key, and type "X" again. Now, *on the same line*, type a colon, followed by "Y+Y1", press [S TO▷], and type "Y" again. When [ENTER] is pressed, the new value of X, x_1 is calculated and stored under the variable name X; subsequently, the new value for Y, y_1, is calculated and stored under the variable name Y. Only the new value of Y is displayed. Pressing [ENTER] a second time results in the calculation of y_2, and you may continue to press [ENTER] until the desired accuracy is obtained.

Copyright © 2012 Pearson Education, Inc. Publishing as Addison-Wesley

Euler's Method and the Sequence Mode of the TI-83 and 83/84 Plus.

While programs for performing Euler's method are included in Part III of this manual, it is also possible to use the sequence mode of the TI-83 and 83/84 Plus to do the calculations. Press [MODE] and move the cursor to the last setting in the fourth line, `Seq`; press [ENTER]. If you now press [Y=], you will see different functions listed there than you have seen before. These functions allow you to define recursive sequences, such as those used to calculate x_i and y_i in Euler's Method.

We must now input a starting value for the subscript, *n*`Min`. Since the initial values are given as x_0 and y_0, type "`0`" and press [ENTER]. We will use `u(`*n*`)` to define the formula for x_{i+1}. For instance, in Example 1 of this section, $x_{i+1} = x_i + .1$; thus we will define `u(`*n*`)=u(`*n*`-1)+.1`. (The letter `u` is obtained by pressing [2nd] [7]; the letter *n* is obtained by pressing [X,T,θ,*n*] while in sequence mode.) Press [ENTER] to move to the next line. Here, we must enter the initial value, `u(0)`, which is the same as x_0; for Example 1, this value is 0. Simply type `0` in this case and press [ENTER]; the calculator will automatically surround this value with brackets. For `v(`*n*`)`, we will enter the corresponding formula for y_{i+1}. In Example 1, this is $y_{i+1} = y_i + (x_i - 2\, x_i\, y_i)(0.1)$. On the calculator, we type "`v(`*n*`-1)+(u(`*n*`-1)-2u(`*n*`-1)v(`*n*`-1))*.1`". (The letter `v` is obtained by pressing [2nd] [8].) Press [ENTER] and input the initial value, y_0, which is 1.5 in this case.

Press [2nd] [WINDOW] to access the `TBLSET` menu and enter "`0`" for `TblStart` and "`1`" for Δ`Tbl`. Press [2nd] [GRAPH] to see the table that represents Euler's Method. (See Figure 1.)

n	u(n)	v(n)
1	.1	1.5
2	.2	1.48
3	.3	1.4408
4	.4	1.3844
5	.5	[illegible]
6	.6	1.2322
7	.7	1.1444

v(n)=1.31360384

Figure 1.

Columns two and three of this table represent the first and second columns of Table 1 in the text.

The graphs can be viewed by first setting the range variables for `Xmin`, `Xmax`, etc., to include the calculated values. You will probably want to turn the function `u(`*n*`)` "off" before pressing [GRAPH].

Copyright © 2012 Pearson Education, Inc. Publishing as Addison-Wesley

Chapter 11 Probability and Calculus

LOCATION IN THE OTHER TEXTS:

Finite Mathematics and Calculus with Applications: Chapter 18

Expected Value and Variance of Continuous Random Variables.

Evaluating Definite Integrals.

Recall from previous discussions that the fnInt function of the TI-83 or 83/84 Plus, as well as the ∫ command on the TI-89, can be used to evaluate definite integrals; review the instructions on page II-40 of this manual, if necessary. The mean and variance in Example 1 of this section of your text can be approximated by using the commands shown in Figures 1 and 2. (Figures 1b and 2b show this problem displayed on the TI-84 Plus with the MathPrint operating system).

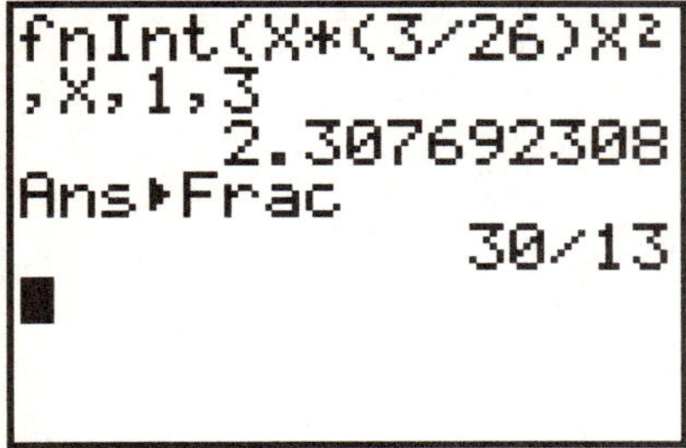

Figure 1a.

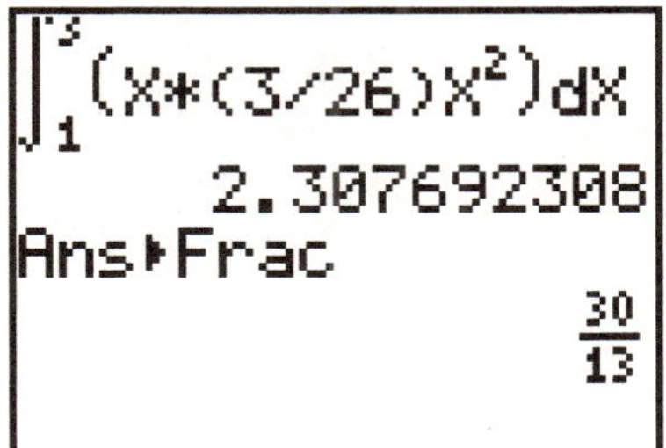

Figure 1b.

Figure 2a.

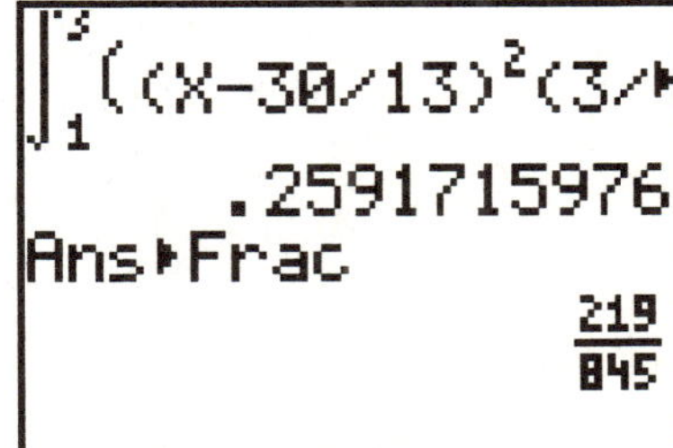

Figure 2b.

The Frac command is the first option in the MATH menu of the TI-83 or 83/84 Plus. Accessing this command immediately after a calculation converts the answer, if possible, to a fraction.

Copyright © 2012 Pearson Education, Inc. Publishing as Addison-Wesley

Special Probability Density Functions.

The Normal Distribution.

Refer to page II-41 of this manual to review how to approximate definite integrals graphically with your calculator. Also, instructions for using the normalcdf command on the TI-83 or 83/84 Plus are located on page II-27 of this manual. Refer to them if necessary for solving Example 3 in your text.

Copyright © 2012 Pearson Education, Inc. Publishing as Addison-Wesley

Chapter 13 The Trigonometric Functions

Definitions of the Trigonometric Functions.

Converting Between Degree and Radian Measure.

To change the angle measure setting on your calculator, press [MODE] on the TI-83 or 83/84 Plus or TI-89. The type of measure for an angle will be shown as either Radian or Degree. Whichever is highlighted is how your calculator will treat any angles used in commands or expressions. To change the setting, move the cursor to the desired angle measure and press [ENTER].

To convert from one angle measure to another, make sure that the calculator is set in the mode to which you wish to convert your angles. In other words, if you want to convert from degrees to radians, set the calculator into Radian mode first; if you want to convert from radians to degrees, set the calculator into Degree mode first. To work Example 1 of the text, set the calculator into Radian mode for part (a). To convert 45° to radians with the TI-83, from the homescreen type "45", then press [2nd] [MATRX] to access the ANGLE menu. Choose option 1, °. The symbol will appear beside "45" on the homescreen; press [ENTER] to perform the conversion. On a TI 83/84 Plus, press [2nd] [APPS] to access the ANGLE menu.

For part (b), first set your calculator to Degree mode. From the homescreen, type "(9π /4)"; press [2nd] [MATRX] (or [2nd] [APPS] on a TI 83/84 Plus) and choose option 3. The symbol ^{r} will appear on the homescreen beside the angle you entered; press [ENTER] to perform the conversion.

On the TI-89, these are located in the Angle collection of commands in the MATH menu.

Sine Regression on the TI-83 and 83/84 Plus and TI-89.

Several exercises in this chapter require the use of the sine regression feature of the TI-83 and 83/84 Plus and TI-89. To use this feature, first save the data to two lists, one for the x-values (in this case, the months), and one for the corresponding y-values (in this case, the average temperature). Review the instructions for entering lists on page II-3 of this manual, if necessary.

To continue with the TI-83 and 83/84 Plus, from the homescreen, press [STAT] and use the right arrow key to move to the CALC submenu. SinReg is option C in this menu; access it and follow it with the name of the x-list, a comma, the name of the y-list, another comma, and a function name for saving the regression equation. Press [ENTER] to perform the calculation.

To complete this problem on the TI-89, from the Data/Matrix Editor, press [F5] and choose SinReg as the Calculation Type.

The result on either calculator is a list of coefficients, a, b, c, and d, for the function y = a*sin(bx+c)+d which best fits the data. You can now set up a statistics plot and graph the scatter plot and the regression equation on the same set of axes.

Copyright © 2012 Pearson Education, Inc. Publishing as Addison-Wesley

Introduction

This part contains hard copy of various programs for graphing calculators. The following calculators are considered:

- Texas Instruments TI-82
- Texas Instruments TI-83 and 83/84 Plus
- Texas Instruments TI-89

Information on how to enter these programs manually and how to run them is provided for each of these calculators at the beginning of each section. However, these short instructions do not claim to tell the whole story on how to operate the calculator; the reader is directed to the Owner's Manuals for the various calculators if more detailed information is desired.

Since the TI-82, TI-83, and TI-83/84 Plus are very similar machines, the same programs will work on either calculator; these calculators are discussed in a single section. The programs for the Texas Instruments calculators are also available via the World Wide Web; see the Texas Instruments sections for more details.

The author of this part wishes to acknowledge the large amount of help he received from two people in the writing of these programs. Thomas Hungerford wrote many of the programs for the Texas Instruments calculators, while Alan Ziv provided valuable expertise and wrote several of the programs for the Casio calculators, which are available upon request from the publisher.

Copyright © 2012 Pearson Education, Inc. Publishing as Addison-Wesley.

Programs for the TI-82, TI-83, and TI-83/84 Plus

Introduction

The following section contains programs for the TI-82, TI-83, and TI-83/84 Plus calculators. These programs allow the calculator to do various financial calculations, linear programming, numerical integration, Euler's method for the numerical solution of differential equations, and to find probabilities associated with the normal distribution.

Entering the Programs Manually: The Program Editor is used to enter a program manually. To access the Program Editor press `PGRM`. To enter a new program, select `NEW` and press `ENTER`; to edit an existing program, select `EDIT`, then move down the list of programs to the program you wish to edit. If you are creating a new program, you will first be prompted for a program name. After entering the name, you will see a colon with a cursor to its left. You are now to enter the first line of the program. Each line of a program begins with a colon; these are supplied by the Editor.

When entering a program from the Program Editor, simple commands may be entered as they appear on the keyboard. Alphabet keys are also easily accessible from the keyboard. Only upper-case letters are available. To enter a letter, press `ALPHA` then the appropriate key. Pressing `2nd` `ALPHA` locks the calculator in alphabet mode. You can easily tell when the calculator is in alphabet mode by the cursor, which flashes an `A` when in alphabet mode. The → symbol which you will need for the programs is the `STO▶` key. In addition to keyboard symbols, you will need to find the commands such as `ClrHome`, `Disp`, and `round` which appear in the programs. These commands are located in various menus; for example `ClrHome` and `Disp` are in the `I/O` submenu of the `PRGM` menu in the Program Editor. So to enter `ClrHome` from the Program Editor, you would press `PGRM`, then select the `I/O` menu, then move down that menu until you reach `8:ClrHome`. Pressing `ENTER` now will place `ClrHome` at the point where you left the Program Editor. Some commands are somewhat difficult to find. On the TI-82, you will need to use the Table of Functions and Instructions in Appendix A of the Guidebook to help you find the proper menus. On the TI-83, there is another way to enter commands into the program. Pressing `2nd` `0` accesses the `CATALOG` menu. Pressing a letter key now will send you to the commands beginning with that letter. Pressing `ENTER` will place the selected command into your program. So to enter `ClrHome` by this method, you would enter the catalog, press `PRGM` to

Copyright © 2012 Pearson Education, Inc. Publishing as Addison-Wesley.

access the letter C (the calculator is already in ALPHA mode), then move down the list until the cursor points at ClrHome, then press [ENTER].

When you are finished entering the program, press [2nd] [MODE] (QUIT) to exit the Program Editor. The program is now stored under the name you have given it.

Entering the Programs from another TI-82, TI-83, or TI-83/84 Plus: Programs and other data may be transferred from a TI-82 to a TI-82 or from a TI-83 or TI-83/84 Plus to a TI-83 or TI-83/84 Plus. It is also possible to transfer files from a TI-82 to a TI-83 or a TI-83/84 Plus, although there are differences between the calculators which could lead to errors in programs transferred in this manner. To perform a transfer you need the communcation link cable which was enclosed with your calculator. Directions on how to perform this process are given in Chapter 16 of the TI-82 Guidebook and in Chapter 19 of the TI-83 Guidebook or TI-83 Plus Guidebook or TI-84-Plus Guidebook.

Entering the Programs from a Computer: The programs in this section are available on the the MyMathLab site for this book. The names of these programs are given at the beginning of each program listing. The programs may be downloaded to a PC then transferred to your calculator. You will need the TI Connect software and cable to perform this transfer. Directions on transferring programs from a PC to the calculator are included with the TI Connectivity Kit.

Running the Programs: To run a program, press [PGRM], make sure that EXEC is selected at the top of the screen, then move down the list of programs to the one you wish to run. Pressing [ENTER] will clear the screen and the name of the program will appear. Pressing [ENTER] again begins execution of the program.

Future Value of an Annuity – FVANN.82P

The following program computes the future value of an annuity given the payment amount, the rate (remember that percents should be converted to decimals), and the number of payments. Payments are made at the end of each period. On the TI-83 or TI-83/84 Plus you may also use the TVM Solver for this calculation. See Chapter 14 of the TI-83 or TI-83/84 Plus Guidebook or pages II-11 through II-16 of this manual for more details.

```
:ClrHome
:Input "PAYMENT?",R
:Input "RATE?",I
:Input "NUMBER OF PYMTS?",N
:Disp "FUTURE VALUE"
:Disp round(R*((1+I)^N-1)/I,2)
```

Copyright © 2012 Pearson Education, Inc. Publishing as Addison-Wesley.

Present Value of an Annuity – PVANN.82P

The following program computes the present value of an annuity given the payment amount, the rate (remember that percents should be converted to decimals), and the number of payments. Payments are made at the end of each period. On the TI-83 or TI-83/84 Plus, you may also use the TVM Solver for this calculation. See Chapter 14 of the TI-83 or TI-83/84 Plus Guidebook or pages II-11 through II-16 of this manual for more details.

```
:ClrHome
:Input "PAYMENT?",R
:Input "RATE?",I
:Input "NUMBER OF PYMTS?",N
:Disp "PRESENT VALUE"
:Disp round(R*(1-(1+I)^(-N))/I,2)
```

Amortization Table – AMORT.82P

The following program asks for the following information about a loan: the starting balance, the rate per period (percents should be converted to decimals), amount of each regular payment, and the number of payments. The program displays an amortization table for this loan in the form of a matrix through which you may scroll by using the arrow keys. In this table the columns are respectively:

Payment Number	Amount of Payment	Interest for the Period	Portion to Principal	New Balance

After scrolling through the table, pressing `ENTER` displays the total payments made and the total interest paid. The program stores the table as matrix `[A]` for future viewing (until the program is run again and a new table takes its place). When the regular payment is too small, the final payment may be quite large. When the regular payment is too large, the final payment may occur before `N` is reached, where `N` is the number of payments entered at the beginning of the program. In this case the matrix will have fewer than `N` rows. Also note that your choice of `N` must be less than or equal to 99.

On the TI-83 or TI-83/84 Plus, you may also use the TVM Solver for this calculation. See Chapter 14 of the TI-83 or TI-83/84 Guidebook or pages II-11 through II-16 of this manual for more details.

```
:ClrHome
:Input "STARTING BALANCE?",B
:Input "RATE PER PERIOD?",I
:Input "PAYMENT?",P
```

Copyright © 2012 Pearson Education, Inc. Publishing as Addison-Wesley.

```
:Input "NUMBER OF PYMTS?",N
:{N,5} → dim([A])
:1 → K
:0 → W
:Lbl 1
:If K=N+1
:Then
:Pause [A]
:ClrHome
:Disp "TOTAL PAYMENTS"
:Disp (J-1)*P+[A](J,2)
:Disp "TOTAL INTEREST"
:Disp W+[A](J,3)
:Stop
:End
:K → [A](K,1)
:P → [A](K,2)
:round(I*B,2) → [A](K,3)
:[A](K,3)+W → W
:P-[A](K,3) → [A](K,4)
:B-[A](K,4) → [A](K,5)
:[A](K,5) → B
:K+1 → K
:If K=N
:Goto 2
:If B≤P
:Goto 2
:Goto 1
:Lbl 2
:K → [A](K,1)
:B+round(I*B,2) → [A](K,2)
:round(I*B,2) → [A](K,3)
:[A](K,2) - [A](K,3) → [A](K,4)
:B-[A](K,4) → [A](K,5)
:{K,5} → dim([A])
:K → J
:N+1 → K
:Goto 1
```

Copyright © 2012 Pearson Education, Inc. Publishing as Addison-Wesley.

Linear Programming – Maximization – SIMPLEX.82P

The following program performs the simplex method on a tableau (matrix) which has been previously input as `[A]`. To run the program you should store your initial tableau as `[A]`. When the program prompts you for the initial matrix, choose `[A]` from the list in the `MATRX` menu. The program will pause to show you intermediate matrices in the calculation, if the calculation takes more than one step. You may use the arrow keys to scroll through this matrix if it is too large for the screen. Press `ENTER` to continue the calculation. The final matrix likewise may be investigated by scrolling. If a solution cannot be found, the calculator reports this fact.

```
:ClrHome
:Disp "INITIAL SIMPLEX"
:Disp "MATRIX"
:Input [A]
:[A] → [B]
:dim [B] → L1
:L1(1) → R
:L1(2) → S
:seq([B](R,I),I,1,(S-1),1) → L2
:min(L2) → T
:If T ≥ 0
:Goto 1
:Lbl X
:1 → J
:Lbl 2
:If [B](R,J)=T
:Goto 3
:J+1 → J
:Goto 2
:Lbl 3
:1 → K
:Lbl 6
:If [B](K,J) ≤ 0
:Goto 4
:[[[B](K,S)/[B](K,J)]] → [E]
:K+1 → K
:Lbl 7
:If [B](K,J) ≤ 0
:Goto 5
:augment([E],[[[B](K,S)/[B](K,J)]]) → [E]
:K+1 → K
```

Copyright © 2012 Pearson Education, Inc. Publishing as Addison-Wesley.

```
:Goto 7
:Lbl 8
:dim [E] → L3
:seq([E](1,I),I,1,L3(2),1) → L4
:min(L4) → M
:1 → I
:Lbl U
:If [B](I,J)=0
:Goto Y
:If [B](I,S)/[B](I,J)=M
:Goto V
:I+1 → I
:Goto U
:Lbl V
:If I=1
:Goto W
:*row([B](I,J)^-1,[B],I) → [B]
:For (K,1,I-1,1)
:*row+(⁻[B](K,J),[B],I,K) → [B]
:End
:For (K,I+1,R,1)
:*row+(⁻[B](K,J),[B],I,K) → [B]
:End
:round([B],9) → [B]
:seq([B](R,I),I,1,(S-1),1) → L2
:min(L2) → T
:If T ≥ 0
:Goto 1
:ClrHome
:Pause [B]►Frac
:Goto X
:Lbl W
:*row([B](I,J)^-1,[B],1) → [B]
:For (K,2,R,1)
:*row+(⁻[B](K,J),[B],I,K) → [B]
:End
:round([B],9) → [B]
:seq([B](R,I),I,1,(S-1),1) → L2
:min(L2) → T
:If T ≥ 0
:Goto 1
```

Copyright © 2012 Pearson Education, Inc. Publishing as Addison-Wesley.

```
:ClrHome
:Pause [B]►Frac
:Goto X
:Lbl 4
:K+1 → K
:If K>(R-1)
:Goto 9
:Goto 6
:Lbl 5
:K+1 → K
:If K>(R-1)
:Goto 8
:Goto 7
:Lbl 1
:ClrHome
:Disp "FINAL SIMPLEX"
:Disp "MATRIX"
:Disp " "
:Pause [B]►Frac
:Stop
:Lbl 9
:Disp "NO MAXIMUM SOLUTION"
:Stop
:Lbl Y
:I+1 → I
:Goto U
```

Trapezoidal Rule – TZOID.82P

The following program uses the Trapezoidal Rule to approximate the value of a definite integral. The function you wish to integrate must be stored in the variable Y_1 before you execute this program. You are also required to input the lower limit of integration (A), the upper limit of integration (B), and the number of subintervals (N).

```
:ClrHome
:Prompt A
:Prompt B
:Prompt N
:(B-A)/N → D
:0 → S
:Y1(A) → S
```

Copyright © 2012 Pearson Education, Inc. Publishing as Addison-Wesley.

```
:Y1(B)+S → S
:For(K,1,N-1,1)
:2Y1(A+K*D)+S → S
:End
:Disp S*D/2
```

Simpson's Rule – SIMPSON.82P

The following program uses Simpson's Rule to approximate the value of a definite integral. The function you wish to integrate must be stored in the variable Y_1 before you execute this program. You are also required to input the lower limit of integration (A), the upper limit of integration (B), and the number of subintervals (N). You must choose an even number for N.

```
:ClrHome
:Prompt A
:Prompt B
:Prompt N
:(B-A)/N → D
:0 → S
:Y1(A) → S
:Y1(B)+S → S
:For(K,1,N/2,1)
:4Y1(A+(2K-1)*D)+S → S
:End
:For(K,1,N/2-1,1)
:2Y1(A+2K*D)+S → S
:End
:Disp S*D/3
```

Integration by Endpoints or Midpoint – LSUM.82P, RSUM.82P, and MSUM.82P

The following programs approximate the value of a definite integral. The function you wish to integrate must be stored in the variable Y_1 before you execute this program. You are also required to input the lower limit of integration (A), the upper limit of integration (B), and the number of subintervals (N).

Left Endpoints – LSUM.82P:

Copyright © 2012 Pearson Education, Inc. Publishing as Addison-Wesley.

```
:ClrHome
:Prompt A
:Prompt B
:Prompt N
:(B-A)/N → D
:0 → S
:For(K,0,N-1,1)
:Y1(A+K*D)+S → S
:End
:Disp S*D
```

Right Endpoints – RSUM.82P:

```
:ClrHome
:Prompt A
:Prompt B
:Prompt N
:(B-A)/N → D
:0 → S
:For(K,1,N,1)
:Y1(A+K*D)+S → S
:End
:Disp S*D
```

Midpoints – MSUM.82P:

```
:ClrHome
:Prompt A
:Prompt B
:Prompt N
:(B-A)/N → D
:0 → S
:For(K,1,N,1)
:Y1(A+(2K-1)*D/2)+S → S
:End
:Disp S*D
```

Copyright © 2012 Pearson Education, Inc. Publishing as Addison-Wesley.

Euler's Method – EULER.82P

The following program performs Euler's method to approximate the solution to the differential equation

$$\frac{dy}{dx} = f(x,y).$$

The function $f(x,y)$ must be stored as `Y1` before you execute this program, and each occurence of y in $f(x,y)$ should be entered as `Y`. You also must enter an initial point (both x and y coordinates), the increment between successive x values, and the value of x at which you wish to find an estimate for y.

```
:ClrHome
:Input "INITIAL X VALUE?",X
:Input "INITIAL Y VALUE?",Y
:Input "INCREMENT?",H
:Input "FINAL X VALUE?",Z
:While X≠Z
:Y+H*Y1 → Y
:X+H → X
:End
:Disp "FINAL (X,Y)"
:Disp X,Y
```

Normal Probability – NRML.82P

The following program computes the probability that a normal random variable with given mean and standard deviation lies between two given values. The probability that the random variable lies below a particular value may be computed by inputting -1E99 for the lower limit; the probability that the random variable lies above a particular value may be computed by inputting 1E99 for the upper limit. The TI-83 or TI-83/84 Plus has a built-in program called `normalcdf` which performs the work of this program. The program `normalcdf` is located in the `DISTR` menu; see the TI-83 or TI-83/84 Plus Guidebook for more details.

```
:ClrHome
:Input "MEAN?",M
:Input "STD DEV?",S
:Input "LOWER LIMIT?",A
:(A-M)/S→A
:If A<⁻10
:⁻10 → A
:If A>10
:10 → A
```

Copyright © 2012 Pearson Education, Inc. Publishing as Addison-Wesley.

```
:Input "UPPER LIMIT?",B
:(B-M)/S→B
:If B<⁻10
:⁻10 → B
:If B>10
:10 → B
:Disp "PR(A<N(M,S)<B)="
:Disp round(fnInt(e^(X^2/⁻2),X,A,B)/√(2*π),4)
```

Copyright © 2012 Pearson Education, Inc. Publishing as Addison-Wesley.

Copyright © 2012 Pearson Education, Inc. Publishing as Addison-Wesley.

Programs for the TI-89

Introduction

The following section contains programs for the TI-89 calculator. These programs allow the calculator to do various financial calculations, linear programming, numerical integration, Euler's method for the numerical solution of differential equations, and to find probabilities associated with the normal distribution.

Entering the Programs Manually: The Program Editor is used to enter a program manually. To access the Program Editor press [APPS] 6. To enter a new program, select `3:New`. You will first be prompted for a program name and for the name of the folder in which you wish to store the program. After entering the name, you will see the name of the program, the commands `Pgrm` and `EndPgrm`, and a colon with a blank line following it. You are to enter the first line of the program on that blank line. Each line of a program begins with a colon; these are supplied by the Editor. When you are done, pressing [2nd] [ESC] (`QUIT`) or [HOME] returns you to the home screen.The program is now stored under the name you have given it.

When entering a program, simple commands may be entered as they appear on the keyboard. Alphabet keys are also easily accessible from the keyboard. To enter an lower case letter, press [alpha] then the appropriate key. To enter a upper case letter, press [⇑] [ALPHA] then the appropriate key. Pressing [alpha] [alpha] or [a-lock] ([2nd] [alpha]) locks the calculator in alphabet mode. You can easily tell when the calculator is in alphabet mode by the icon `a` which appears in the status line at the bottom of the calculator screen. The → symbol which you will need for these programs is the [STO▶] key. In addition to keyboard symbols, you will need to find the commands such as `ClrIO`, `Disp`, and `round` which appear in the programs. These commands are located in various menus. For example, `round` is in the `MATH Number` menu, which is accessed by by pressing [2nd] [5], then selecting `1:Number`. Pressing the right arrow key exposes a long list of commands. Moving down the list to `3:round(` and pressing [ENTER] will place `round(` at the point where you left the Program Editor. The menus for some commands are somewhat difficult to find; all commands are listed in the catalog, which is accessed by pressing [CATALOG]. Pressing a letter key now will send you to the commands beginning with that letter, and pressing [ENTER] will place the selected command into your program. So to enter `ClrIO` by this method, you would enter the catalog, press [)] to enter a `c` (the calculator is

Copyright © 2012 Pearson Education, Inc. Publishing as Addison-Wesley.

already in alphabet mode), move down the list until the cursor points at `ClrIO`, then press ENTER.

To edit an existing program, access the Program Editor and select `2:Open`. With the Variable box selected, press the right arrow key to see a list of programs in the current folder. Move the cursor down to the name of the program you wish to edit, then press ENTER twice. The programs are listed in alphabetical order; if you don't see the program you want, continue using the down arrow to expose more of the list.

Entering the Programs from another TI-89: Programs and other data may be transferred from a TI-89 to another TI-89. To perform a transfer you need the communication link cable which was enclosed with your calculator. Directions on how to perform this process are given in Chapter 22 of the TI-89 Guidebook.

Entering the Programs from a Computer: The programs in this section are available on the the MyMathLab site for this book. The names of these programs are given at the beginning of each program listing. The programs may be downloaded to your computer and then transferred to your calculator. You will need the TI Connect software and cable to perform this transfer. Directions on transferring programs from a PC to the calculator are included with TI Connectivity Kit.

Running the Programs: To run a program, enter the name of the program on the entry line of the Home screen. Pressing ENTER begins execution of the program. For example, entering `fvann()` on the entry line, then pressing ENTER would start the first program listed below. You can also locate the program using the VAR-LINK feature on the calculator (2nd –) – see Chapter 21 of the TI-89 Guidebook for details.

Future Value of an Annuity – FVANN.89P

The following program computes the future value of an annuity given the payment amount, the rate (remember that percents should be converted to decimals), and the number of payments. Payments are made at the end of each period.

```
:fvann()
:Prgm
:ClrIO
:Input "Payment?",r
:Input "Rate?",i
:Input "Number of payments?",n
:Disp "Future value"
:Disp round(r*((1+i)^n-1)/i,2)
:EndPrgm
```

Copyright © 2012 Pearson Education, Inc. Publishing as Addison-Wesley.

Present Value of an Annuity – PVANN.89P

The following program computes the present value of an annuity given the payment amount, the rate (remember that percents should be converted to decimals), and the number of payments. Payments are made at the end of each period.

```
:pvann()
:Prgm
:ClrIO
:Input "Payment?",r
:Input "Rate?",i
:Input "Number of payments?",n
:Disp "Present value"
:Disp round(r*(1-(1+i)^(-n))/i,2)
:EndPrgm
```

Amortization Table – AMORT.89P

The following program asks for the following information about a loan: the starting balance, the rate per period (percents should be converted to decimals), amount of each regular payment, and the number of payments. The program creates an amortization table for this loan, and displays the total payments made and the total interest paid over the life of the loan. In the amortization table the columns are respectively:

Payment Number	Amount of Payment	Interest for the Period	Portion to Principal	New Balance

The program stores the amortization table as matrix `a` for future viewing (until the program is run again and a new table takes its place). To view the table, open the Data/Matrix Editor (`APPS` 6) and select `1:Current`. When the regular payment is too small, the final payment may be quite large. When the regular payment is too large, the final payment may occur before `n` is reached, where `n` is the number of payments entered at the beginning of the program. In this case the matrix will have fewer than `n` rows. Also note that your choice of `n` must be less than or equal to 999.

```
:amort()
:Prgm
:ClrIO
:Input "Starting balance?",b
:Input "Rate per period?",i
:Input "Payment?",p
```

Copyright © 2012 Pearson Education, Inc. Publishing as Addison-Wesley.

```
:Input "Number of payments?",n
:1 → k
:0 → w
:{} → col1
:{} → col2
:{} → col3
:{} → col4
:{} → col5
:Lbl lbl1
:If k=n+1 Then
:NewData a,col1,col2,col3,col4,col5
:ClrIO
:Disp "Total Payments"
:Disp (j-1)*p+col2[j]
:Disp "Total Interest"
:Disp w+col3[j]
:Stop
:EndIf
:k → col1[k]
:p → col2[k]
:round(i*b,2) → col3[k]
:col3[k]+w → w
:p-col3[k] → col4[k]
:b-col4[k] → col5[k]
:col5[k] → b
:k+1 → k
:If k=n
:Goto lbl2
:If b≤p
:Goto lbl2
:Goto lbl1
:Lbl lbl2
:k → col1[k]
:b+round(i*b,2) → col2[k]
:round(i*b,2) → col3[k]
:col2[k]-col3[k] → col4[k]
:b-col4[k] → col5[k]
:k → j
:n+1 → k
:Goto lbl1
:EndPrgm
```

Copyright © 2012 Pearson Education, Inc. Publishing as Addison-Wesley.

Linear Programming – Maximization – SIMPLEX.89P

The following program performs the simplex method on a tableau (matrix) which you provide. When the program prompts you for the initial matrix, you may input it at the prompt. If you have previously stored the matrix, simply enter the variable name at the prompt. The program will pause to show you intermediate matrices in the calculation, if the calculation takes more than one step. You may use the cursor pad to scroll through this matrix if it is too large for the screen. Press ENTER to continue the calculation. The final matrix likewise may be investigated by scrolling. If a solution cannot be found, the calculator reports this fact.

```
:simplex()
:Prgm
:ClrIO
:Disp "Initial simplex matrix"
:Input a
:a → b
:dim(b) → l1
:l1[1] → r
:l1[2] → s
:seq(b[r,i],i,1,s-1,1) → l2
:min(l2) → t
:If t ≥ 0
:Goto lbl1
:Lbl lbl13
:1 → j
:Lbl lbl2
:If b[r,j]=t
:Goto lbl3
:j+1 → j
:Goto lbl2
:Lbl lbl3
:1 → k
:Lbl lbl6
:If b[k,j] ≤ 0
:Goto lbl4
:[[b[k,s]/(b[k,j])]] → e
:k+1 → k
:Lbl lbl7
:If b[k,j] ≤ 0
```

Copyright © 2012 Pearson Education, Inc. Publishing as Addison-Wesley.

```
:Goto lbl5
:augment(e,[[b[k,s]/(b[k,j])]]) → e
:k+1 → k
:Goto lbl7
:Lbl lbl8
:dim(e) → l3
:seq(e[1,i],i,1,l3[2],1) → l4
:min(l4) → M
:1 → i
:Lbl lbl10
:If b[i,j]=0
:Goto lbl14
:If b[i,s]/(b[i,j])=m
:Goto lbl11
:i+1 → i
:Goto lbl10
:Lbl lbl11
:If i=1
:Goto lbl12
:mRow(b[i,j]^(⁻1),b,i) → b
:For k,1,i-1,1
:mRowAdd(⁻b[k,j],b,i,k) → b
:EndFor
:For k,i+1,r,1
:mRowAdd(⁻b[k,j],b,i,k) → b
:EndFor
:seq(b[r,i],i,1,s-1,1) → l2
:min(l2) → t
:If t ≥ 0
:Goto lbl1
:ClrIO
:Pause b
:Goto lbl13
:Lbl lbl12
:mRow(b[i,j]^(⁻1),b,1) → b
:For k,2,r,1
:mRowAdd(⁻b[k,j],b,i,k) → b
:EndFor
:seq(b[r,i],i,1,s-1,1) → l2
:min(l2) → t
:If t ≥ 0
```

Copyright © 2012 Pearson Education, Inc. Publishing as Addison-Wesley.

```
:Goto lbl1
:ClrIO
:Pause b
:Goto lbl13
:Lbl lbl4
:k+1 → k
:If k>r-1
:Goto lbl9
:Goto lbl6
:Lbl lbl5
:k+1 → k
:If k>r-1
:Goto lbl8
:Goto lbl7
:Lbl lbl1
:ClrIO
:Disp "Final Simplex Matrix"
:Disp " "
:Pause b
:Stop
:Lbl lbl9
:Disp "No Maximum Solution"
:Stop
:Lbl lbl14
:i+1 → i
:Goto lbl10
:EndPrgm
```

Trapezoidal Rule – TZOID.89P

The following program uses the Trapezoidal Rule to approximate the value of a definite integral. The function you wish to integrate must be stored in the variable y_1 before you execute this program. You are also required to input the lower limit of integration (a), the upper limit of integration (b), and the number of subintervals (n).

```
:tzoid()
:Prgm
:ClrIO
:Prompt a
:Prompt b
```

Copyright © 2012 Pearson Education, Inc. Publishing as Addison-Wesley.

```
:Prompt n
:(b-a)/n → d
:0 → s
:a → x
:y1(x) → s
:b → x
:y1(x)+s → s
:For k,1,n-1,1
:a+k*d → x
:2*y1(x)+s → s
:EndFor
:Disp s*d/(2.)
:EndPrgm
```

Simpson's Rule – SIMPSON.89P

The following program uses Simpson's Rule to approximate the value of a definite integral. The function you wish to integrate must be stored in the variable y_1 before you execute this program. You are also required to input the lower limit of integration (a), the upper limit of integration (b), and the number of subintervals (n). You must choose an even number for n.

```
:simpson()
:Prgm
:ClrIO
:Prompt a
:Prompt b
:Prompt n
:(b-a)/n → d
:0 → s
:a → x
:y1(x) → s
:b → x
:y1(x)+s → s
:For k,1,n/2,1
:a+(2*k-1)*d → x
:4*y1(x)+s → s
:EndFor
:For k,1,n/2-1,1
:a+2*k*d → x
:2*y1(x)+s → s
```

Copyright © 2012 Pearson Education, Inc. Publishing as Addison-Wesley.

```
:EndFor
:Disp s*d/(3.)
:EndPrgm
```

Integration by Endpoints or Midpoint – LSUM.89P, RSUM.89P, MSUM.89P

The following programs approximate the value of a definite integral. The function you wish to integrate must be stored in the variable y_1 before you execute this program. You are also required to input the lower limit of integration (a), the upper limit of integration (b), and the number of subintervals (n).

Left Endpoints – LSUM.89P:

```
:lsum()
:Prgm
:ClrIO
:Prompt a
:Prompt b
:Prompt n
:(b-a)/n → d
:0 → s
:For k,0,n-1,1
:a+k*d → x
:y1(x)+s → s
:EndFor
:Disp s*d
:EndPrgm
```

Right Endpoints – RSUM.89P:

```
:rsum()
:Prgm
:ClrIO
:Prompt a
:Prompt b
:Prompt n
:(b-a)/n → d
:0 → s
:For k,1,n,1
```

Copyright © 2012 Pearson Education, Inc. Publishing as Addison-Wesley.

```
:a+k*d → x
:y1(x)+s → s
:EndFor
:Disp s*d
:EndPrgm
```

Midpoints – MSUM.89P:

```
:msum()
:Prgm
:ClrIO
:Prompt a
:Prompt b
:Prompt n
:(b-a)/n → d
:0 → s
:For k,1,n,1
:a+(2*k-1)*d/2 → x
:y1(x)+s → s
:EndFor
:Disp s*d
:EndPrgm
```

Euler's Method – EULER.89P

The following program performs Euler's method to approximate the solution to the differential equation

$$\frac{dy}{dx} = f(x,y).$$

The function $f(x,y)$ must be stored as y_1 before you execute this program, and each occurence of y in $f(x,y)$ should be entered as `y`. You also must enter an initial point (both x and y coordinates), the increment between successive x values, and the value of x at which you wish to find an estimate for y.

```
:euler()
:Prgm
:ClrIO
:Input "Initial x value?",x
:Input "Initial y value?",y
:Input "Increment?",h
```

Copyright © 2012 Pearson Education, Inc. Publishing as Addison-Wesley.

```
:Input "Final x value?",z
:While x≠z
:y+h*y1(x) → y
:x+h → x
:EndWhile
:Disp "Final (x,y)"
:Disp x,y
:EndPrgm
```

Normal Probability – NRML.89P

The following program computes the probability that a normal random variable with given mean and standard deviation lies between two given values. The probability that the random variable lies below a particular value may be computed by inputting $-\infty$ for the lower limit; the probability that the random variable lies above a particular value may be computed by inputting ∞ for the upper limit. The infinity sign can be found on the keyboard as [♦] [CATALOG].

```
:nrml()
:Prgm
:ClrIO
:Input "Mean?",m
:Input "Standard deviation?",s
:Input "Lower limit?",a
:(a-m)/s→a
:Input "Upper limit?",b
:(b-m)/s→b
:Disp "Pr(a<N(m,s)<b)="
:Disp nInt(e^(⁻x^2/2),x,a,b)/√(2*π))
:EndPrgm
```

Copyright © 2012 Pearson Education, Inc. Publishing as Addison-Wesley.

Part IV

General Instructions for Excel

In this part, we give an overview of the general concepts and tools that involve spreadsheets. The spreadsheet software we use is Microsoft's Excel, but the principles should hold for any current software package.

Copyright © 2012 Pearson Education, Inc. Publishing as Addison-Wesley

Spreadsheets

Introduction

One could argue that the advent of spreadsheet and word-processing software ushered in the PC as a primary tool in every workplace. In today's world, wherever there is a table of data, it is stored in a spreadsheet. Some of the things one can do with data via spreadsheets are:

- Data Arrangement
- Calculation
- Visualization
- Database Management
- Statistical Analysis
- Predictions
- Optimization

Spreadsheet packages also come with powerful built-in programming languages, making their versatility almost limitless. The best part about spreadsheets is that the initial learning curve is very short! We use Microsoft Excel 2007 for all demonstrations in this manual, but due to the standardization in today's spreadsheets, one should be able to apply this material to almost any spreadsheet package. Throughout this manual, spreadsheet and PC terminology will be used. In Figure 1, the typical window one sees is displayed with many of the names we use in this manual.

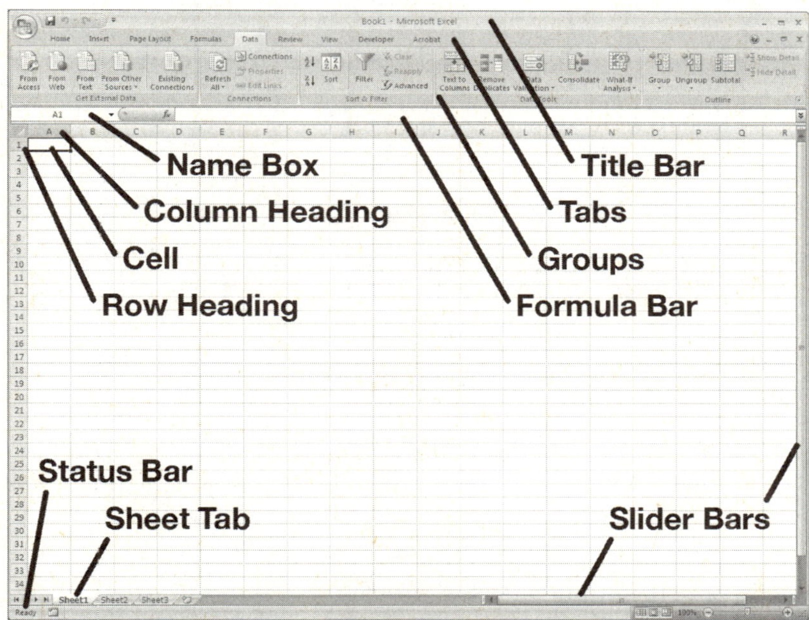

Figure 1: These names should help keep you straight.

Copyright © 2012 Pearson Education, Inc. Publishing as Addison-Wesley

Files

Collections of sheets are stored in files called **workbooks**. When you open Excel, it starts with a brand new workbook called "Book1.xlsx". Figure 1 is very close to what you will see. Notice the title of the workbook in the **title bar** at the top of the window. If you want to open a previously saved workbook, you can do so through the **Office button** in the upper left hand or open it directly by finding the file and double-clicking it.

Each workbook consists of sheets, accessed via the **sheet tabs** at the bottom of the screen (Figure 1). Each sheet can be either a **chart** or a **worksheet**. Charts are any kind of visualization of data, such as scatter plots or bar graphs. Worksheets are the "spreadsheets", the place where we (and the program) do all of the work.

Cells

Everything begins with the **cell**. If you can understand exactly what a cell is and what properties it has, the rest should be relatively easy. *It is crucial that you understand this section!*

A cell is one of the many rectangles you see on a worksheet. One or more cells are **selected** when they are surrounded by a bold outline with a **fill handle** in the lower right hand corner of the outline (we will discuss the purpose of the fill handle later). You can select a single cell by clicking on it or select a rectangular set of them by click-dragging (Figure 2)

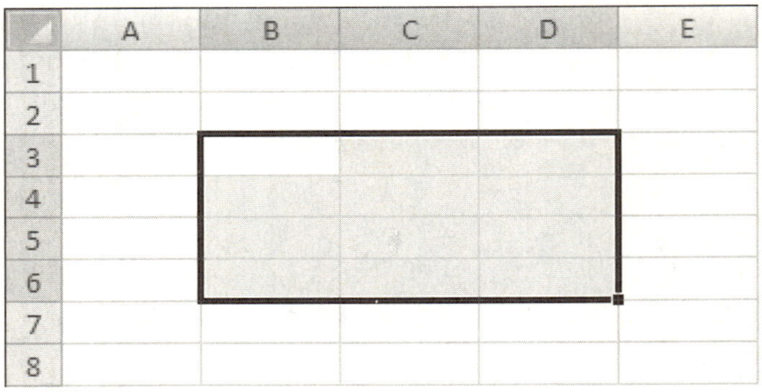

Figure 2: The range `B3:D6` is selected.

The cell's primary function is to hold and display "stuff." *The cell has four major attributes, which you must learn:*

- Address
- Format
- Content
- Value

Cell Address

The **address** of a cell is indicated by the row and column in which it sits. (If you have ever played Battleship®, you know exactly how this works!) The **column heading** is always one or two letters (ranging from `A` to `IV`) and the **row heading** is always a number (from `1` to `65536`). A cell's address consists of the column letter(s) and the row number. For example, if a

Copyright © 2012 Pearson Education, Inc. Publishing as Addison-Wesley

cell sits in column `B` and row `3`, then its address is `B3` (Figure 3). Note: The address of the selected cell is given in the **name box**.

Figure 3: The address of the selected cell is `B3`

When a rectangular **range** of cells is selected, the address of the range is given by the address of the upper left cell and the lower right cell, separated by a colon. In Figure 2 the upper left cell of the selected range is `B3` and the lower right cell is `D6`, thus the range is denoted by `B3:D6`. Note: When a range of cells forms a single column or row, the two end cells are used to denote the range.

Cell Format

A cell can contain many things, like numbers and text, but how the cell displays those things can vary greatly. Suppose you type into every cell of a range (say `B1:B5`) the number 0.75. Without any formatting, the cells all display 0.75. By right-clicking a cell, you bring up a plethora of options, including **Format Cells...** (like most everything else, this can also be obtained through the tabs). The first tab you get in the **Format Cells** pop-up window is **Number** (Figure 4). Here is where you can format the cell to display its value as anything from dates to currency. You can even dictate how many decimal places the cell will display (which can cause Excel to round the displayed value but *not* the actual value). In Figure 5 you can see some of the possibilities, where all the cells contain the same value 0.75.

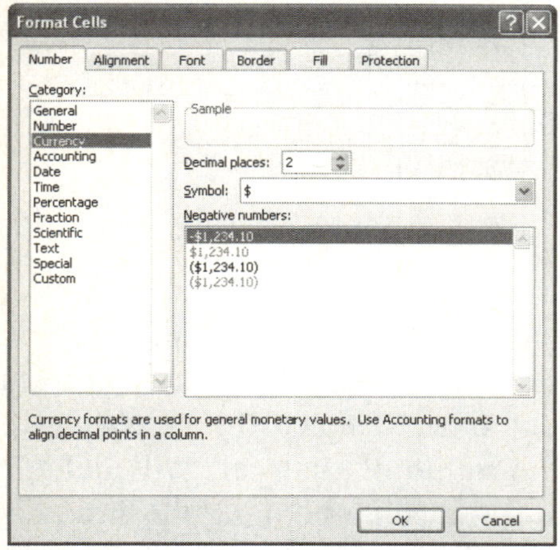

Figure 4: Format Cells pop-up

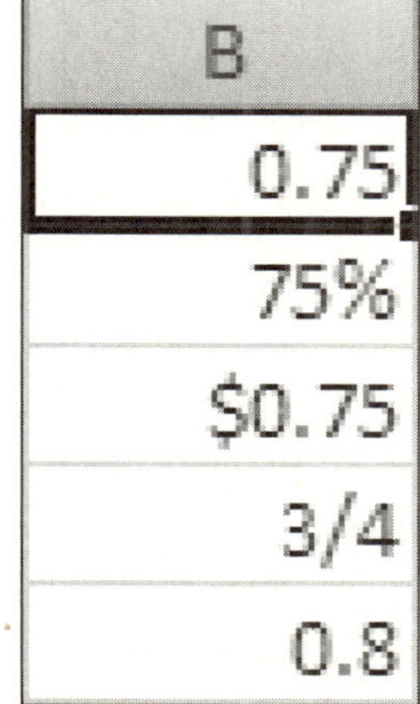

Figure 5: Assorted formatting of 0.75.

Copyright © 2012 Pearson Education, Inc. Publishing as Addison-Wesley

Cell Content

The **content** of a cell is whatever you type into it, which is usually a number, text, or a formula. The content of the cell is not necessarily what the cell displays! In other words, you may type a formula into a cell and press `Enter`, then what the cell *displays* is the value of the formula, while the *content* of the cell is the formula. Whenever you select a cell, the content of the cell is shown in the **formula bar**. For example suppose you select a cell, type the formula `=3/4`, and press `Enter`. If you formatted the cell as a percentage, then the cell will display the value 75%. Now reselect cell, look in the formula bar and you should see the formula `=3/4` (Figure 6).

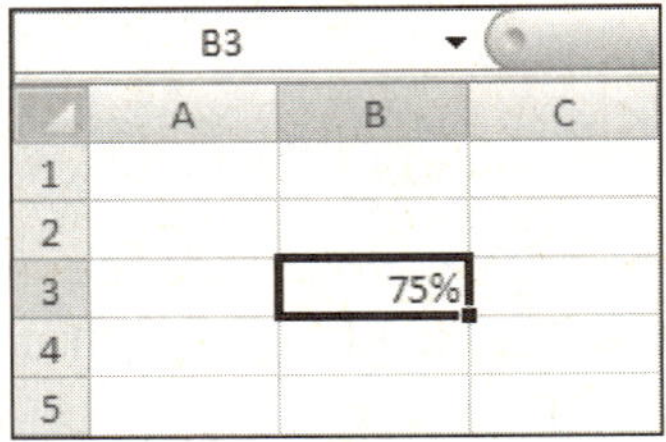

Figure 6: Cell `B3` displays 75%, has a value of 0.75, and contains the formula `=3/4`.

This is what makes spreadsheets so powerful. You will be able to enter formulas into cells that use values from other cells. In turn, these other cells may contain formulas themselves!

There is something very important to remember when using longer, more complicated formulas in Excel. *Excel does not correctly obey the standard order of operations!* Recall from basic algebra that $-3^2 = -9^2$ and $(-3)^2 = 9$, because an exponent applies only to whatever is to its immediate left. Unfortunately, Excel is programmed to apply negation (the negative sign) before applying exponents, whether parentheses are present or not. So the formula =–A1^2 will return the value of –A1 squared, not the negative of A1 squared. Therefore, when using Excel, you should adopt the convention of enclosing the base and the exponent in parentheses. Table 1 shows the result of three different formulas. If you wish to calculate -3^2, then you should use the formula shown in the last column of the table.

	A	B	C	D
1				
2	A2	=A3^2	=-A3^2	=-(A3^2)
3	3	9	9	-9

Table 1.

Cell Value

Every cell has a numeric **value**, even an empty one (which has a default value of 0). If you type a number into a cell, then that will also be the cell's value. If you type a numeric formula into a cell, then that cell's value will be the outcome of the formula. When the cell contains text or has a formula that returns text, then the value of that cell will default to 0. Remember, the value of a cell, what the cell displays, and the cell's content are three different things. As seen in the example illustrated in Figure 6, the value of the cell is a 0.75, while the cell's content is the formula `=3/4` and the cell's display is 75%.

Copyright © 2012 Pearson Education, Inc. Publishing as Addison-Wesley

Formulas

As seen earlier, you can type formulas into a cell. All formulas start with an "=". A formula can have numbers, algebraic operations, **functions**, and addresses for cells and ranges. Notice that we do not have variables! Any time we want to use a value from another cell in a formula, we can use the cell's address instead. Let us consider an example.

Suppose you wanted to store in cell `D2` the average of three numbers that you have stored in cells `B1`, `B2` and `B3`. This can be done with a formula. In cell `D2`, type `=(B1+B2+B3)/3` and press `Enter`. The cell `D2` displays the average. If you reselect cell `D2`, the formula bar will display the formula that the cell contains, while the cell displays its value (Figure 7).

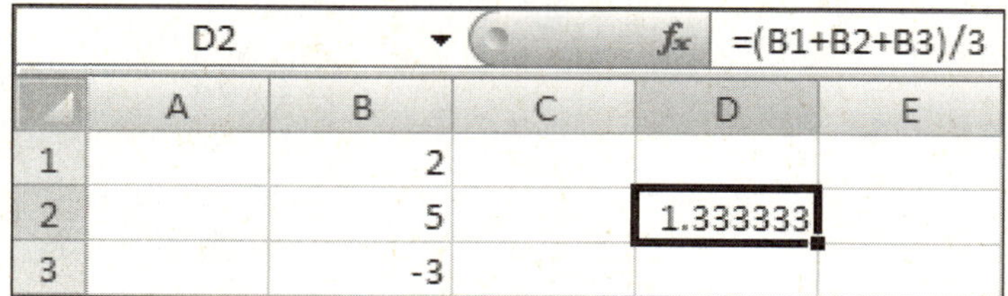

Figure 7: Average of cells via a simple formula.

For many common calculations (such as averaging) there are built-in functions. To view a list of these functions, click on a cell and type "=". The name box changes and becomes a drop-down list of functions you have used recently, but right now the most interesting choice is **More Functions...** (Figure 8). Through this choice, you can explore and use a whole host of functions; it is certainly worth your while to check this out! You can also type in functions directly if you know the correct spelling and appropriate arguments. In Figure 9, the function `=AVERAGE(B1:B3)` gives the same result as seen in Figure 7.

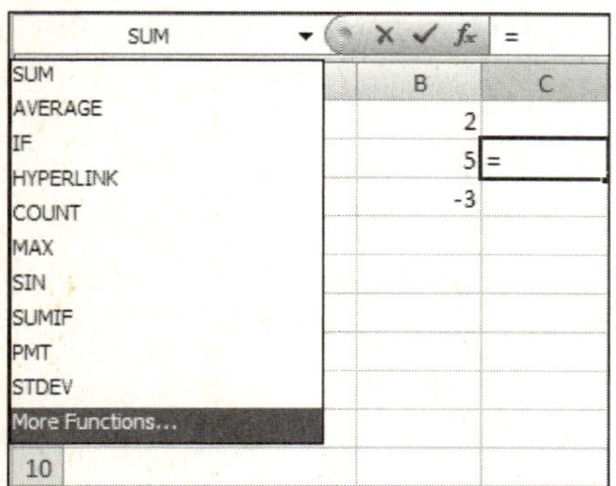

Figure 8: Average of cells via a simple formula.

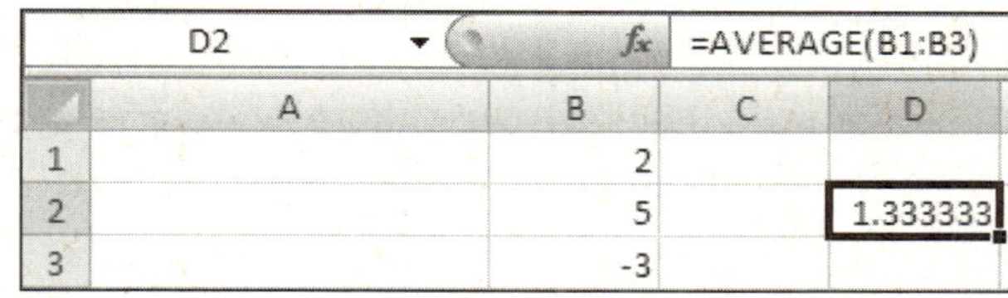

Figure 9: Average of cells via a simple formula.

Copy, Paste, and Fill

Anyone who has used a PC is probably quite familiar with cut, copy, and paste. When copying cells, if the cell's content is anything other than a formula, the contents are identically transcribed. However, when the content of a copied cell is a function, then the addresses within the function are adjusted by the relative distance from the copied cell to the pasted cell.

Copyright © 2012 Pearson Education, Inc. Publishing as Addison-Wesley

As an example, consider Figure 10 where you have a table of values of which you want the average of each column. Type in the appropriate formula for averaging the first column, copy that cell and paste into the cell under the next column (Figure 11). Notice how the address range A1:A3 is shifted by exactly one column value to B1:B3. This is precisely the relative change from the copied cell (A4) to the pasted cell (B4). If you pasted from the cell A4 to the cell C9, the relative change would be two to the left and five down, thus the pasted formula would be =AVERAGE(C6:C8).

A4 | fx =AVERAGE(A1:A3)

	A	B	C	D	E
1	3	2	5	8	
2	7	5	-9	8	
3	2	-3	2	3	
4	4				

Figure 10: The first average function is typed in.

B4 | fx =AVERAGE(B1:B3)

	A	B	C	D	E
1	3	2	5	8	
2	7	5	-9	8	
3	2	-3	2	3	
4	4	1.333333			

Figure 11: The next average is copied from the first.

B4 | fx =AVERAGE(B1:B3)

	A	B	C	D	E
1	3	2	5	8	
2	7	5	-9	8	
3	2	-3	2	3	
4	4	1.333333			

Figure 12: The fill handle is grabbed and dragged.

B4 | fx =AVERAGE(B1:B3)

	A	B	C	D	E
1	3	2	5	8	
2	7	5	-9	8	
3	2	-3	2	3	
4	4	1.333333	-0.66667	6.333333	

Figure 13: All of the averages are correctly filled in.

Coming back to our example, there is a much quicker way of copying one cell to multiple cells. The technique is called **filling**, and with most things, there are several ways of accomplishing it. As noted earlier, every highlighted cell has a fill handle in the lower right-hand corner. If you hold-click this handle (grab) and drag in any direction, an automatic copy/paste is performed on all covered cells. In Figure 12 the handle on cell B4 is shown grabbed and dragged across cells C4 and D4. When the mouse is released, the average function is pasted into the both C4 and D4 with the appropriate adjustment to the address ranges within the function (Figure 13).

There may be times when you will copy/paste a cell and you want one or more parts of the address ranges in the function's arguments to *not* change. The trick is to use a $ in front of every part of the addresses you want to freeze. For instance, suppose you are filling from a cell with the formula =A1+B3-C7. If you do not want the column of the address A1 to change, you would instead use the formula =$A1+B3-C7. If you do not want the row address of B3 to change, you would use the formula =A1+B$3-C7. If you do not want the address C7 to change at all, you would use the formula =A1+B3-$C$7. When an address is completely frozen, e.g. $C$7, this is called an **absolute reference**. When no part of the cell reference is frozen, e.g. C7, this is called a **relative reference**. When part of a cell reference is frozen, e.g. $C7 or C$7, this is called a **mixed reference**.

Copyright © 2012 Pearson Education, Inc. Publishing as Addison-Wesley

Charts

A picture is worth a thousand words. Spreadsheets are loaded with graphical tools for displaying sets of data. The best part is how easy it is to create a **chart** of data. Consider the data in Figure 14 of homework and test grades over the first four chapters of a class. Select the data to be visualized (range `B1:C5`) and via the **Insert** tab, select the option for a **2-D Column** chart under the **Column** icon in the **Charts** group. As seen in Figure 15, the **Charts** group gives many choices of possible charts, including bar charts, pie charts, and scatter plots. Click on the expand option at the bottom right of the **Charts** group to open the window shown in Figure 15 displaying all of the chart types.

	A	B	C
1	Chapter	HW	Test
2	1	75	73
3	2	82	78
4	3	78	80
5	4	91	88

Figure 14: These are class grades to be visualized.

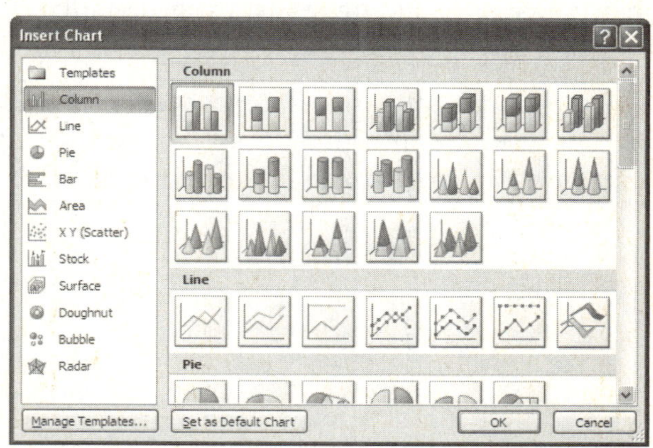

Figure 15: There are a host of choices in charts.

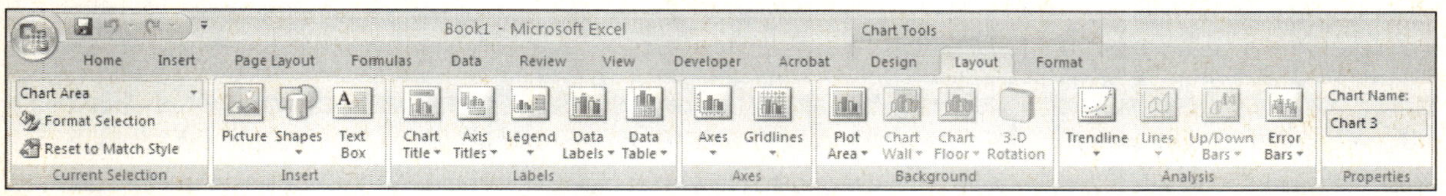

Figure 16: The layout tab gives lots of control over the chart.

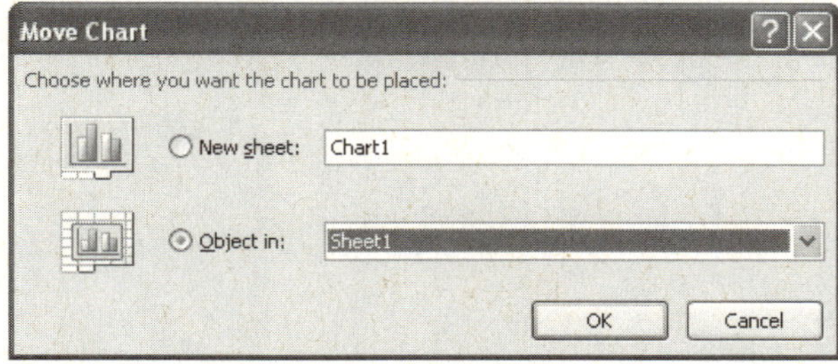

Figure 17: The chart can be added to the data spreadsheet or placed on a separate sheet.

Once you have selected the appropriate chart type, the chart will appear within your Excel worksheet. Selecting the chart will bring up three new tabs; **Design, Layout, and Format** (Figure 16**)**. These tabs give a vast number of choices to fine tune and annotate your graph. By selecting the **Move Chart** icon in the **Location** group of the **Design** tab you may move your chart to its own sheet or embed it in another (Figure 17).

Copyright © 2012 Pearson Education, Inc. Publishing as Addison-Wesley

Conclusion

Once you become comfortable with the spreadsheet environment, you should experiment with all the available charts, formulas, and built-in abilities. The possibilities for effectively and efficiently using Excel, or any other spreadsheet program, are practically endless. If you are using Microsoft Word or Microsoft PowerPoint, all tables and charts from Excel can be copied and pasted into a Word document or PowerPoint presentation so that you can create attractive papers and presentations. You can even import data from other programs and databases, as well as from the World Wide Web. Even though this manual only demonstrates how to use Excel for problems that arise in finite mathematics and in applied calculus, Excel also has applicability to other courses, including statistics, accounting, economics, biology, chemistry, and physics, just to name a few.

You may also find spreadsheets useful for keeping track of your college credits, personal finances, even as an address book for e-mail and mailing addresses and phone numbers. You are limited only by your imagination and patience.

Copyright © 2012 Pearson Education, Inc. Publishing as Addison-Wesley

Part V

Detailed Instructions for Excel

Part V contains detailed steps and instructions for using Microsoft's Excel to complete many examples from *Finite Mathematics*, tenth edition, *Calculus with Applications*, tenth edition, *Calculus with Applications: Brief Version*, tenth edition, and *Finite Mathematics and Calculus with Applications*, ninth edition. Most of the instructions provided can be applied to other current spreadsheet programs.

Copyright © 2012 Pearson Education, Inc. Publishing as Addison-Wesley

Detailed Instructions for Finite Mathematics

This section of Part V contains detailed instructions for using Microsoft Excel for *Finite Mathematics*, tenth edition, and related chapters in *Finite Mathematics and Calculus with Applications*, ninth edition. Chapter 1 is common to these two texts, as well as *Calculus with Applications*, tenth edition, and *Calculus with Applications: Brief Version*, tenth edition. This section is organized by chapters in *Finite Mathematics*; since not all chapters require detailed explanations of spreadsheet use, some chapters are not mentioned here.

In this manual, section titles from the textbooks are indicated in italics. References are made to specific examples and exercises from the corresponding sections of each chapter, so you should have your textbook nearby as you read through these instructions.

Copyright © 2012 Pearson Education, Inc. Publishing as Addison-Wesley

Chapter 1 Linear Functions

LOCATION IN THE OTHER TEXTS:

Calculus with Applications: Chapter 1
Calculus with Applications, Brief Version: Chapter 1
Finite Mathematics and Calculus with Applications: Chapter 1

Linear Functions and Applications.

Using *Goal Seek* to Solve Equations.

If an equation contains only one variable, then the **Goal Seek** function can solve it. Begin by creating a table containing the relevant variable and formulas. For instance, to solve Example 6(a) of your text with Excel, beginning in cell A1, set up appropriate column headings to represent the units (x), revenue, cost, and profit. Cell A2 will serve as the variable cell, holding the place of x in our formulas. You need not enter anything in this cell. In the other cells in row 2, type the following:

Cell	Contents
B2	=24*A2
C2	=20*A2+100
D2	=B2-C2

The table should appear similar to Table 1.

	A	B	C	D
1	**Units (x)**	**R=24x**	**C=20x+100**	**Profit=R-C**
2	0	0	100	-100

Table 1.

Access the **Goal Seek** function under the **What-If Analysis** drop-down menu in the **Data Tools** group of the **Data** tab. For the cell to "set," type D2; for the value to set this cell to, type 0; the cell to be "changed" is the cell acting as our variable, which is cell A2. The **Goal Seek** pop-up should look like Figure 1. (Note: **Goal Seek** will automatically change all cell references entered into absolute references; see Part I of this manual for more information on absolute references.)

Copyright © 2012 Pearson Education, Inc. Publishing as Addison-Wesley

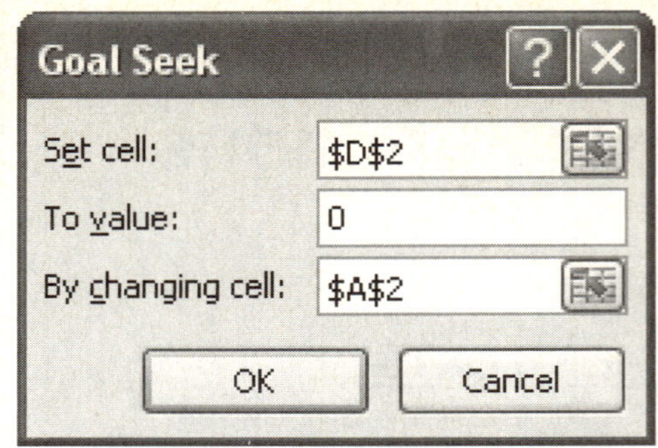

Figure 1.

Click **OK** and cell `A2` (the number of units) will be adjusted until cell `D2` (the profit) is 0. We see in Figure 2 that 25 units must be sold for this company to break even.

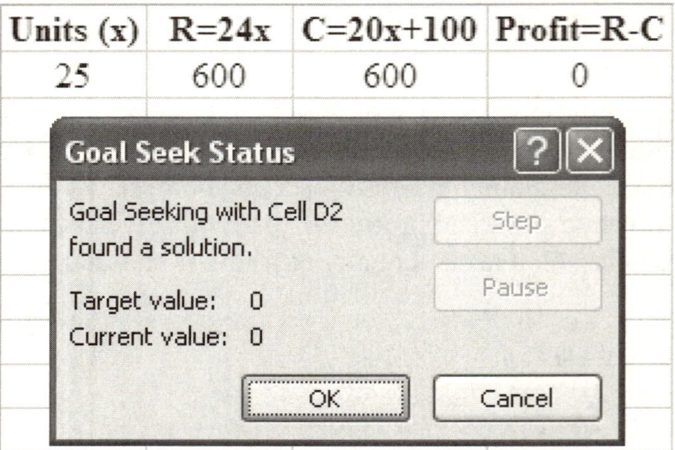

Figure 2.

The Least Squares Line

Finding the Least Squares Regression Line.

To complete the example in this section, begin by entering the lists of data points into columns `A` and `B`. We can now use the functions **SLOPE**, **INTERCEPT**, and **CORREL** to find *m*, *b*, and *r*, respectively. Type the following into the indicated cells:

<u>Cell</u>	<u>Contents</u>
`E1`	`=SLOPE(B2:B11,A2:A11)`
`E3`	`=INTERCEPT(B2:B11,A2:A11)`
`E5`	`=CORREL(B2:B11,A2:A11)`

When using the **SLOPE** and **INTERCEPT** functions, the list containing the *y*-values must be first, as shown above. The final result should resemble Table 2 on the next page.

Copyright © 2012 Pearson Education, Inc. Publishing as Addison-Wesley

	A	B	C	D	E
1	**Year**	**Death Rate**		**Slope =**	-0.55969697
2	10	84.4		**y-intercept =**	90.33333333
3	20	71.2		**r =**	-0.96290058
4	30	80.5			
5	40	73.4			
6	50	60.3			
7	60	52.1			
8	70	56.2			
9	80	46.5			
10	90	36.9			
11	100	34			

Table 2.

Displaying Data and the Regression Line on the Same Graph.

Begin by entering the lists of data points and creating an **XY (Scatter)** chart in the spreadsheet. (See Part I of this manual for additional assistance with creating charts.) Right-click on any data point in the chart and select **Add Trendline**. Select the second type, **Linear**. Check the boxes in front of "Display equation on chart" and "Display R-squared value on chart." Click **Close** and the least squares line and the resulting value of r^2 are displayed. You may need to click and drag the text box containing the equation and r^2 value to move it to a more convenient position within the chart. The final result should resemble Figure 3.

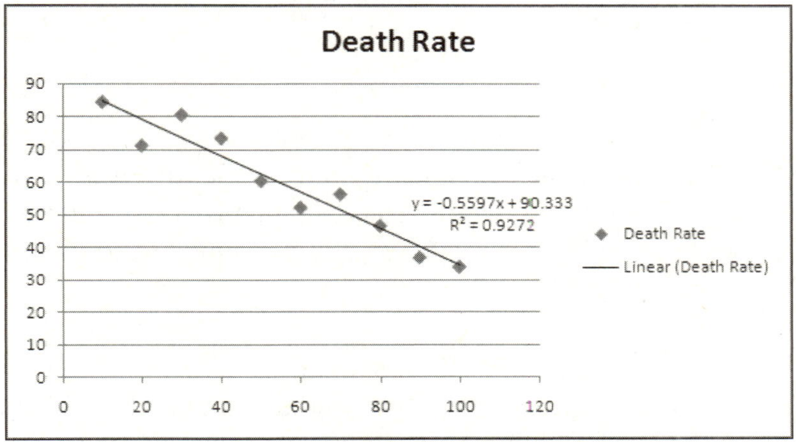

Figure 3.

It is important to remember that this method gives you r^2, which is the square of the correlation coefficient. To find the value of r, you must take the square root of this value, or use the CORREL function as described previously.

Copyright © 2012 Pearson Education, Inc. Publishing as Addison-Wesley

Chapter 2 Systems of Linear Equations and Matrices

LOCATION IN THE OTHER TEXTS:

Finite Mathematics and Calculus with Applications: Chapter 2

Solution of Linear Systems by the Gauss-Jordan Method.

Row Operations.

As indicated in your text, a clever use of the **Copy** and **Paste Special** commands can minimize the arithmetic required by the Gauss-Jordan Method for solving systems of linear equations. Begin by setting up the matrix, with appropriate row and column headings, in the cell range `A1:E4`, as shown in Table 1.

	A	B	C	D	E
1		**x**	**y**	**z**	**Constant**
2	**R1**	1	-1	5	-6
3	**R2**	3	3	-1	10
4	**R3**	1	3	2	5

Table 1.

Highlight the entire matrix, including the row and column headings, select **Copy** from the **Clipboard** group of the **Home** tab, and click on cell `A6`. Select **Paste** from the **Clipboard** group of the **Home** tab to paste the entire matrix. Since the first row operation we are asked to perform is $-3R_1 + R_2 \rightarrow R_2$, highlight the entire second row of the copied matrix and type in the following:

Cell	Contents
`B8`	`=-3*B2:E2+B3:E3`

Do not press **Enter***!* Instead, press **Ctrl-Shift-Enter** (all three keys simultaneously). This will cause each of the cells in our copied R_2 to be updated with the values resulting from the row operation. To complete the next row operation, $-R_1 + R_3 \rightarrow R_3$, highlight the entire third row of the copied matrix and type in the following, pressing **Ctrl-Shift-Enter** when finished:

Cell	Contents
`B9`	`=-B2:E2+B4:E4`

Copyright © 2012 Pearson Education, Inc. Publishing as Addison-Wesley

The new matrix, with the updated rows, should look like Table 2.

	A	B	C	D	E
1		x	y	z	**Constant**
2	**R1**	1	-1	5	-6
3	**R2**	0	6	-16	28
4	**R3**	0	4	-3	11

Table 2.

Now, highlight the entire new matrix, and select **Copy** from the **Clipboard** group of the **Home** tab. Click on cell `A11`, and this time, select **Paste Special** from the **Paste** drop-down menu. In the window that pops up, select **Values** and click **OK**. This will copy only the values, not the formulas, from the previous matrix operations. To perform the next row operation from this example in your text, highlight the first row of the third matrix, type `=B8:E8+6*B7:E7`, and press **Ctrl-Shift-Enter** to update the new row. Notice that we are now creating our formulas from the second matrix, the one that was updated. To finish with the third matrix, select the third row, type `=2*B8:E8-3*B9:E9` and press **Ctrl-Shift-Enter**. This matrix should now look like Table 3.

	A	B	C	D	E
1		x	y	z	**Constant**
2	**R1**	6	0	14	-8
3	**R2**	0	6	-16	28
4	**R3**	0	0	-23	23

Table 3.

Continue by copying and using the **Paste Special** command to create the fourth matrix, and make sure you create the formulas by referencing the previously completed matrix, until the process is complete.

The Solver.

Excel's built-in **Solver** can find solutions to many different types of systems of equations, using a method similar to one you will learn in Chapter 4 of your textbook. Among the types of systems **Solver** can solve are systems of linear equations with unique solutions. The **Solver** should be located in the **Data Tools** group of the **Data** tab in Excel. If it is *not*, you will need to locate your original installation disk, select **Add-Ins**, and install the **Solver** program.

You should take care to set up a worksheet that contains as much helpful information as possible, including headings for rows, columns, and even individual cells, so that you can keep track of what all the entries represent. To complete Example 2 from your text using **Solver**, we can set up a worksheet similar to that in Table 4, with the following formulas in the indicated cells:

Copyright © 2012 Pearson Education, Inc. Publishing as Addison-Wesley

Cell	Contents
`B3`	`=B1-D1+5*F1`
`B4`	`=3*B1+3*D1-F1`
`B5`	`=B1+3*D1+2*F1`

	A	B	C	D	E
1	**x=**		**y=**		**z=**
2					
3	x-y+5z=-6	0	=	-6	
4	3x+3y-z=10	0	=	10	
5	x+3y+2z=5	0	=	5	

Table 4.

From the **Data Tools** group of the **Data** tab, select **Solver**. Type in the address of the cell containing the first formula, in this case, cell `B3`, as the "Target Cell," then check "Value of" and type in the value of the constant for the first equation, which is -6 in this example. The cells we wish to change are the cells next to the labels for *x*, *y*, and *z*, so type `B1,D1,F1` into the box below "By Changing Cells." Our constraints here are the equations themselves, and we have already typed these in. Select **Add**, and type in the name of the cell containing the first formula. Select "=" from the drop-down menu, then type in the name of the cell containing the constant from the first equation. If your worksheet is set up like Table 4, then the **Add** window should appear as in Figure 1. If so, click **Add**.

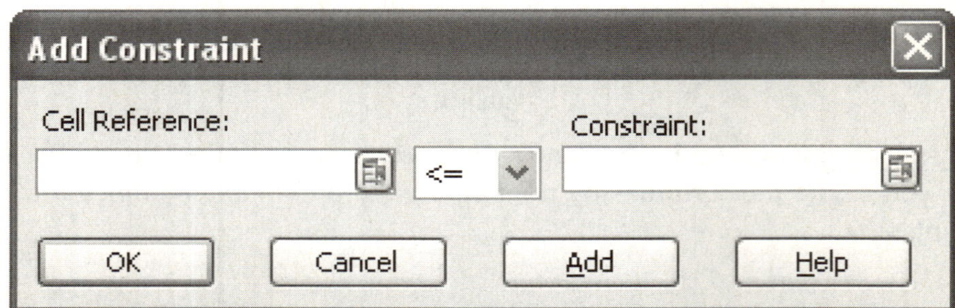

Figure 1.

Now, we need to add the second equation as a constraint, so type in B4 as the "Cell Reference," select "=" from the drop-down menu, and enter D4 as the "Constraint." We need to add one more equation, so click **Add** again and repeat this process for the third equation. When it is entered, click **OK**. This will return you to the **Solver** window, which should now look like Figure 2 on the next page.

Copyright © 2012 Pearson Education, Inc. Publishing as Addison-Wesley

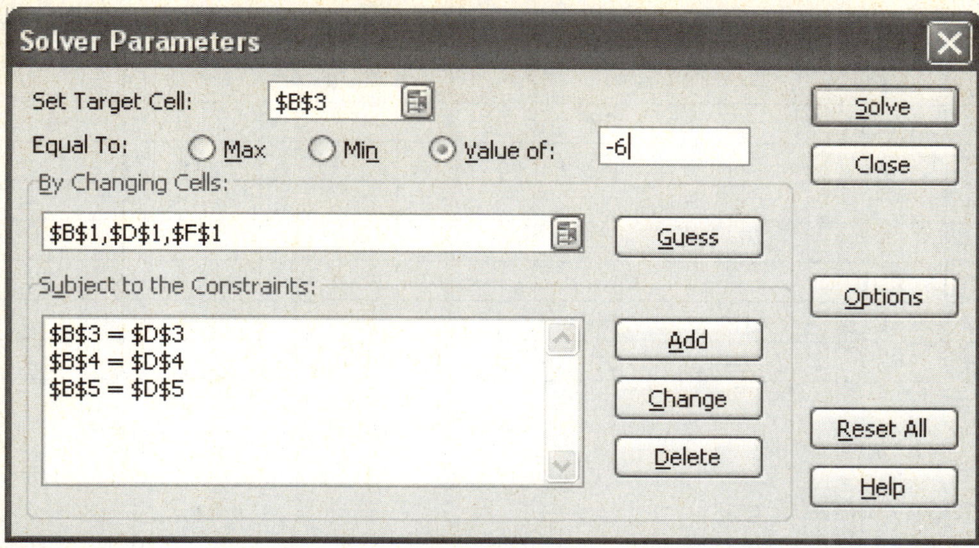

Figure 2.

Click **Solve**. The program will now adjust the values of the variables until all constraints are satisfied, if possible. The top row of the worksheet should now contain the solution to the system of equations, $x = 1$, $y = 2$, and $z = -1$.

The **Solver** cannot solve systems of equations that have no solution, nor can it solve systems with infinitely many solutions. If you try to use it to solve one of these types of systems, a new window will appear letting you know that **Solver** was unable to complete the problem.

Addition and Subtraction of Matrices.

Example 7 of this section of your text can be completed with Excel by first entering the matrices, C and K, then using a **Ctrl-Shift-Enter** trick as we did with the matrix row operations. This technique is highly useful if there are large matrices that need to be added or subtracted.

To complete Example 7, enter the matrices C and K into a worksheet, in a manner similar to Table 5. Then, highlight a block of empty cells that is the *same size as the answer should be*. In this case, you need to highlight a block of cells that is two rows by three columns. While the empty block is highlighted, type `=A2:C3-E2:G3`. This indicates that the first matrix is located in the cell range `A2:C3` and we wish to subtract from it the second matrix, which is located in the range `E2:G3`. Press **Ctrl-Shift-Enter** and the empty block will be filled with the resulting matrix. (See Table 5 on the next page.)

Copyright © 2012 Pearson Education, Inc. Publishing as Addison-Wesley

	A	B	C	D	E	F	G
1	**C**				**K**		
2	22	25	38		5	10	8
3	31	34	35		11	14	15
4							
5							
6	**C-K**						
7	17	15	30				
8	20	20	20				

Table 5.

Multiplication of Matrices.

The **MMULT** function in Excel can be used to multiply matrices. To complete Example 5 of this section of your text, enter the two matrices to be multiplied, *A* and *B*, and highlight a block of empty cells that is the same size as *AB* should be. In this case, we need to highlight a block of cells that is three rows by three columns. Type `=MMULT(A2:B4,D2:F3)` and press **Ctrl-Shift-Enter** to see the resulting product. Similarly, the product *BA* can be calculated by highlighting a set of empty cells that is 2 rows by 2 columns, typing `=MMULT(D2:F3,A2:B4)`, and pressing **Ctrl-Shift-Enter**. (See Table 6.)

	A	B	C	D	E	F
1	**A**			**B**		
2	1	-3		1	0	-1
3	7	2		3	1	4
4	-2	5				
5						
6	**AB**				**BA**	
7	-8	-3	-13		3	-8
8	13	2	1		2	13
9	13	5	22			

Table 6.

Matrix Inverses.

Once a square matrix has been entered into a worksheet, its inverse can be calculated with the **MINVERSE** function. To complete Example 2 from this section of your text, enter matrix *A*, then highlight a set of empty cells that is the same size as the inverse should be, which is in this case 3 rows by 3 columns. Type `=MINVERSE(A2:C4)` and press **Ctrl-Shift-Enter** to calculate the inverse. (See Table 7.)

Copyright © 2012 Pearson Education, Inc. Publishing as Addison-Wesley

	A	B	C	D	E	F	G
1	**A**				**A-inverse**		
2	1	0	1		0	0	0.33333
3	2	-2	-1		-0.5	-0.5	0.5
4	3	0	0		1	0	-0.33333

Table 7.

A Note About Round-off Errors and Matrix Inverses.

Occasionally, when performing calculations involving matrix inverses, one or more entries in the resulting matrix may look like `1E-10`, or something similar. Keep in mind that this represents 1 times 10 to the -10th power, or 0.0000000001. A result like this in a problem that originally contained no numbers of this form is due to round-off error during Excel's calculations. Usually, it is safe to treat any similar results as "0."

Copyright © 2012 Pearson Education, Inc. Publishing as Addison-Wesley

Chapter 4 Linear Programming: The Simplex Method

LOCATION IN THE OTHER TEXTS:

Finite Mathematics and Calculus with Applications: Chapter 4

Maximization Problems.

Excel's **Solver** can quickly find solutions to linear programming problems of all types, when a single solution exists. You will again need to carefully set up a worksheet so that the results of running the **Solver** are easy to interpret. Table 1 is a convenient way to set up Example 1 from this section of your text.

	A	B	C	D
1	**x1=**		**x2=**	
2				
3	**Maximize**	z=	0	
4				
5	**Subject to**			
6	x1 + x2 <=65	0	<=	65
7	12x1 + 15x2 <= 900	0	<=	900

Table 1.

The following formulas were entered into the indicated cells:

Cell	Contents
C3	=25*B1+30*D1
B6	=B1+D1
B7	=12*B1+15*D1

Select **Solver** from the **Data Tools** group of the **Data** tab. The "Target" cell is the one containing the expression to be maximized, which, in this case, is cell `C3`. Make sure that "Max" is selected here since this is a maximization problem. The cells to be changed are the cells next to the variable labels, which in Table 1 are cells `B1` and `D1`. Add the two constraints as described in our previous discussion of **Solver**. Click **Options** and check "Assume non-negative," then click **OK**. Now click **Solve**, and the solution is calculated.

Notice that, when adding constraints, the drop-down menu allows you to use "=", "≥", and "≤". Thus, minimization problems as well as non-standard problems with single solutions, all discussed later in Chapter 4 of your text, can be completed using **Solver**. Simply choose the appropriate symbol when adding constraints into **Solver**.

Copyright © 2012 Pearson Education, Inc. Publishing as Addison-Wesley

Chapter 5 Mathematics of Finance

LOCATION IN THE OTHER TEXTS:

Finite Mathematics and Calculus with Applications: Chapter 5

Simple and Compound Interest.

Comparing Interest Schemes.

A spreadsheet can be used to compare two interest schemes, as shown in Figure 2 of this section of your text. To create a similar spreadsheet, enter appropriate column headings to represent years, interest compounded annually, and simple interest. (See Table 1.) Fill column `A` with the numbers `1` through `20` to indicate the years since the principal was invested, then type in the following:

Cell	Contents
`B2`	`=1000*(1+0.1)^A2`
`C2`	`=1000*(1+0.1*A2)`

Drag and fill columns `B` and `C` to complete the table. (Only the first 11 rows of the resulting table are shown here.)

	A	B	C
1	**Years**	**Interest Compounded Annually**	**Simple Interest**
2	1	$1,100.00	$1,100.00
3	2	$1,210.00	$1,200.00
4	3	$1,331.00	$1,300.00
5	4	$1,464.10	$1,400.00
6	5	$1,610.51	$1,500.00
7	6	$1,771.56	$1,600.00
8	7	$1,948.72	$1,700.00
9	8	$2,143.59	$1,800.00
10	9	$2,357.95	$1,900.00
11	10	$2,593.74	$2,000.00

Table 1.

Copyright © 2012 Pearson Education, Inc. Publishing as Addison-Wesley

Effective Rate.

Excel has many built-in functions for calculating financial values. The function **EFFECT** calculates the effective interest rate corresponding to a given nominal interest and a given number of payments per year. To complete Example 7 from this section of your text, type the following in any blank cell in a worksheet:

Cell	Contents
Any cell	`=EFFECT(.06,2)`

Present Value.

The **PV** function calculates the present value of an account, when the nominal interest rate, total number of interest payments, and the future value of the account are known. Enter the interest rate, as a decimal; the total number of interest payments over the life of the account; the amount that will be paid into the account (by the account holder) at the end of each payment period; then the desired future value of the account followed by the account type. In any empty cell, type the following to calculate the present value:

Cell	Contents
Any cell	`=PV(.062,5,0,6000,1)`

Notice that `1` was entered as the "Type" of account; this indicates that the account holder will make a payment at the beginning of a period. However, in this example, the account holder will be making only one payment, not periodic payments; this is why `0` was entered for "Pmt." When you press **Enter**, the resulting present value will be entered into the selected cell. This value will be shown as a *negative* value, appearing in red, since this amount must be paid out by the account holder.

Copyright © 2012 Pearson Education, Inc. Publishing as Addison-Wesley

Future Value of an Annuity.

Ordinary Annuities.

Excel has a built-in function, **FV**, for calculating the future value of an ordinary annuity. As with present value, this function is for when the nominal interest rate, total number of interest payments, and the future value of the account are known. In any empty cell, type the following to calculate the future value in Example 3 from this section of your text:

Cell	Contents
Any cell	`=FV(.06,7,-22000,0,0)`

Notice that, this time, `0` was selected as the "Type"; this is because Example 4 clearly states that money will be deposited at the *end* of each year. The payment, "Pmt", is entered as a *negative* number since this amount will be paid out by the account holder at the end of each payment period. Also note that for an ordinary annuity, `0` should be entered as the present value.

If, as in Example 5, interest is compounding a certain number of times per year, this should be reflected in the "Rate" entered. For instance, in Example 4(a), 7.2% interest is compounded monthly, so we would enter 0.072/12 as the "Rate."

Cell	Contents
Any cell	`=FV(.072/12,12*20,-200,0,0)`

Present Value of an Annuity; Amortization.

Amortization Schedules.

To create an amortization schedule with Excel, similar to the one in Figure 14 from this section of your text, begin by entering appropriate column headings. (See Table 2.) Fill column `A` with the numbers `0` through `12`, indicating the payment numbers for this example. You may wish, at this point, to format the cells in columns `B`, `D`, and `E` so that numerical values are displayed as currency. (See Part I of this manual for information on formatting cells.) In cell `E2`, enter the amount of the loan, $1000. Since monthly payments are $88.85, fill cells `B3:B14` with this amount. Since interest is 1% per month (12% compounded monthly), and payments are applied to interest before principal, enter the following formulas into the indicated cells:

Cell	Contents
`C3`	`=0.01*E2`
`D3`	`=B3-C3`
`E3`	`=E2-D3`

Copyright © 2012 Pearson Education, Inc. Publishing as Addison-Wesley

Drag and fill the rest of columns C, D and E. Since the last entry is negative, the final payment should probably be adjusted by the lender to $88.83, so that the customer avoids overpaying the loan.

	A	B	C	D	E
1	**Payment Number**	**Amount of Payment**	**Interest for Period**	**Portion to Principal**	**Principal at End of Period**
2	0				$1,000.00
3	1	$88.85	$10.00	$78.85	$921.15
4	2	$88.85	$9.21	$79.64	$841.51
5	3	$88.85	$8.42	$80.43	$761.08
6	4	$88.85	$7.61	$81.24	$679.84
7	5	$88.85	$6.80	$82.05	$597.79
8	6	$88.85	$5.98	$82.87	$514.91
9	7	$88.85	$5.15	$83.70	$431.21
10	8	$88.85	$4.31	$84.54	$346.67
11	9	$88.85	$3.47	$85.38	$261.29
12	10	$88.85	$2.61	$86.24	$175.05
13	11	$88.85	$1.75	$87.10	$87.96
14	12	$88.85	$0.88	$87.97	-$0.02

Table 2.

Copyright © 2012 Pearson Education, Inc. Publishing as Addison-Wesley

Chapter 8 Counting Principles; Further Probability Topics

LOCATION IN THE OTHER TEXTS:

Finite Mathematics and Calculus with Applications: Chapter 8

The Multiplication Principle; Permutations.

Factorial.

Excel's command for calculating factorials is the **FACT** function. To calculate 5!, type the following:

Cell	Contents
Any cell	`=FACT(5)`

Permutations.

The permutation function is **PERMUT**. To calculate P(10,3), type the following in any empty cell:

Cell	Contents
Any cell	`=PERMUT(10,3)`

Combinations.

To calculate $\binom{8}{3}$ with Excel, we use the **COMBIN** function. In any empty cell, enter the following:

Cell	Contents
Any cell	`=COMBIN(8,3)`

Copyright © 2012 Pearson Education, Inc. Publishing as Addison-Wesley

Detailed Instructions for Calculus with Applications

This section of Part V contains detailed instructions for using Microsoft Excel for *Calculus with Applications*, tenth edition, *Calculus with Applications: Brief Version*, tenth edition, and *Finite Mathematics and Calculus with Applications*, ninth edition. (Instructions for Chapter 1 of all three texts begin on page V-3.) The section is organized by chapters in *Calculus with Applications*; since not all chapters require detailed explanations of spreadsheet use, some chapters are not mentioned here.

In this manual, section titles from the textbooks are indicated in italics. References are made to specific examples and exercises from the corresponding sections of each chapter, so you should have your textbook nearby as you read through these instructions.

Copyright © 2012 Pearson Education, Inc. Publishing as Addison-Wesley

Chapter 2 Nonlinear Functions

LOCATION IN THE OTHER TEXTS:

Calculus with Applications, Brief Version: Chapter 2
Finite Mathematics and Calculus with Applications: Chapter 10

Quadratic Functions.

Quadratic Regression.

In Exercise 63 of this section of your text, a quadratic function is fit to a set of data representing the median age of women at their first marriage. The quadratic regression feature of Excel can be used to complete this exercise. There are two ways to do this, but the easiest by far is to begin by creating a scatter plot of the data set. Begin by entering appropriate column headings for the year and the median age. (See Table 1.)

	A	B
1	Year (Since 1900)	Age
2	40	21.5
3	50	20.3
4	60	20.3
5	70	20.8
6	80	22.0
7	90	23.9
8	100	25.1

Table 1.

Enter the data, using the numbers 40 through 100 to represent the years 1940 through 2000. Highlight the data set, including column headings, and select the **Scatter** icon in the **Charts** group of the **Insert** tab. Right-click on any data point in the chart and select **Add trendline**. Select the **Polynomial** type, and make sure that "2" is selected as the "order" since we wish to fit a second degree polynomial to the data. Select the "Display equation on chart" option. Click **Close**. The quadratic model, and its equation, will be displayed on the chart. The result should appear similar to Figure 1 on the next page. (You may wish to drag and move the equation box to keep it from interfering with the rest of your chart.)

Copyright © 2012 Pearson Education, Inc. Publishing as Addison-Wesley

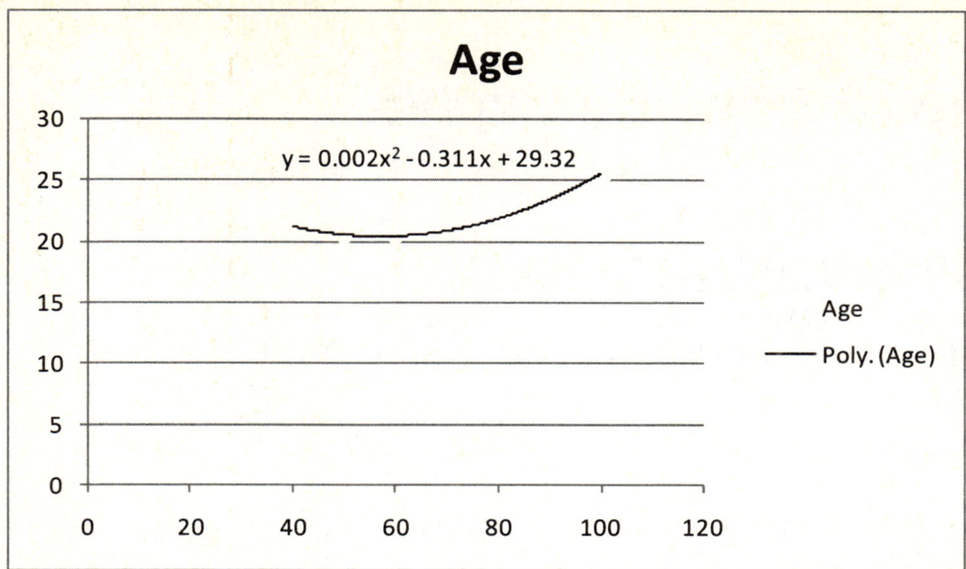

Figure 1.

This polynomial trendline feature of Excel can be used to fit up to a sixth degree polynomial model to a set of data. To do so, simply change the "order" to the degree of the polynomial you wish to fit.

Exponential Functions.

Exponential Regression.

Example 7(b) of this section of the text asks us to find an exponential function that models the given corn production data. For simplicity, subtract 1930 from each year, so that 1930 becomes year 0, 1940 becomes year 10, and so on. After entering appropriate column headings, enter the years and the corresponding corn production levels in adjacent columns. Follow the instructions given above for creating a scatter plot and adding a trendline. For the trendline type, select **Exponential.**

Applications: Growth and Decay; Mathematics of Finance.

Power Regression.

In Exercise 108(d) from the review exercises of your textbook, you are asked to use power regression to fit an appropriate model to a set of data. Following the same procedures as outlined above, use the **Power** type when adding a trendline.

Copyright © 2012 Pearson Education, Inc. Publishing as Addison-Wesley

Chapter 3 The Derivative

LOCATION IN THE OTHER TEXTS:

Calculus with Applications, Brief Version: Chapter 3
Finite Mathematics and Calculus with Applications: Chapter 11

Limits.

Creating Limit Tables.

As indicated in the discussion following Example 1, limits of functions can be estimated by observing appropriate values of a function in a table. To use Excel to estimate the limit in this example, begin by choosing appropriate column headings and entering, in column A, a list of x-values that surround $x = 2$. (See Table 1.) Enter the following contents into cell B2 and drag to fill the rest of column B:

Cell	Contents
B2	=(A2^2+4)/(A2-2)

	A	B
1	x	(x^2+4)/(x-2)
2	1.9	-76.1
3	1.99	-796.01
4	1.999	-7996.001
5	1.9999	-79996.0001
6	2.0001	80004.0001
7	2.001	8004.001
8	2.01	804.01
9	2.1	84.1

Table 1.

Copyright © 2012 Pearson Education, Inc. Publishing as Addison-Wesley

Rates of Change.

Estimating Instantaneous Rates of Change.

To generate the table in Figure 26 from this section of your text, choose appropriate column headings, like those in Table 2, for the value of h, the value of $f(x+h)$, the value of $f(x)$, and the value of the difference quotient. Enter the following values into the appropriate cells:

Cell	Contents
A2	`1`
B2	`=37.791*1.021^(10+A2)`
C2	`=37.791*1.021^00`
D2	`=(B2-C2)/A2`

Enter additional values of h in column A, making sure that each is smaller than the previous one, and then drag and fill columns B, C and D to create a table similar to Table 2.

	A	B	C	D
1	**h**	**f(x+h)**	**f(x)**	**Inst. Rate of Change**
2	1	47.49758701	46.52065329	0.976933719
3	0.1	46.61743556	46.52065329	0.967822642
4	0.01	46.53032247	46.52065329	0.966917771
5	0.001	46.52162012	46.52065329	0.966827346
6	0.0001	46.52074998	46.52065329	0.966818305

Table 2.

It should be clear from observing the last column of Table 2 that, as h approaches 0, the instantaneous rate of change of $f(x)$ is approximately 0.96682 when $x = 10$.

Copyright © 2012 Pearson Education, Inc. Publishing as Addison-Wesley

Chapter 7 Integration

LOCATION IN THE OTHER TEXTS:

Calculus with Applications, Brief Version: Chapter 7
Finite Mathematics and Calculus with Applications: Chapter 15

Area and Definite Integrals.

Summation.

To use Excel to complete Example 2 from this section of your text, type in appropriate column headings similar to those in Table 1. In column A, type in the numbers 1 through 4 to indicate the number of the rectangle. In column B, type in the x-values representing the midpoints of the bases of the rectangles. Enter the following contents into the indicated cells:

Cell	Contents	
C2	=2*B2	(evaluate $f(x)$ at B2)
F1	1	(the value of Δx)

Drag and fill the rest of column C. Since we now wish to sum the values $f(x_i)\Delta x$, we can do this with a **SUM** command. In cell C6, type =SUM(C2:C5)*$F$1. The result should look similar to Table 1.

	A	B	C	D	E	F
1	**i**	**xi**	**f(xi)**		**Delta-x**	1
2	1	0.5	1			
3	2	1.5	3			
4	3	2.5	5			
5	4	3.5	7			
6		**Sum:**	16			

Table 1.

Copyright © 2012 Pearson Education, Inc. Publishing as Addison-Wesley

Numerical Integration.

The Trapezoidal Rule.

As indicated in the discussion after Example 1 in your text, the trapezoidal rule can be done quickly with Excel. Begin by creating column headings for the i, x_i, and $f(x_i)$. To complete Example 1, fill column `A` with the integer values `0` through `4`. Next, enter the following contents into the indicated cells:

Cell	Contents	
`B2`	`=$F$1+A2*$F$2`	
`C2`	`=SQRT(B2^2+1)`	(evaluate the function)
`F1`	`0`	(the left endpoint for the definite integral)
`F2`	`0.5`	(the value of Δx)

Notice the absolute cell references for F1 and F2, since you will want the same left endpoint and Δx throughout. Highlight cells `B2` and `C2`, then drag to fill columns `B` and `C`. Finally, to apply the trapezoidal rule, in cell `F3`, type `=F2*(0.5*C2+SUM(C3:C5)+0.5*C6)`. The result should appear similar to Table 2 below.

	A	B	C	D	E	F
1	**i**	**xi**	**f(xi)**		**Left endpoint**	0
2	0	0	1		**Delta-x**	0.5
3	1	0.5	1.11803		**Approximation**	2.97653
4	2	1	1.41421			
5	3	1.5	1.80278			
6	4	2	2.23607			

Table 2.

Simpson's Rule.

To complete Example 2 via Simpson's rule on Excel, set up a table exactly like the one you set up for the trapezoidal rule. In cell `F3`, which will contain the Simpson's rule approximation for the integral, type `=(F2/3)*(C2+4*C3+2*C4+4*C5+C6)`. The result should appear similar to Table 3 below.

	A	B	C	D	E	F
1	**i**	**xi**	**f(xi)**		**Left endpoint**	0
2	0	0	1		**Delta-x**	0.5
3	1	0.5	1.11803		**Approximation**	2.97653
4	2	1	1.41421			
5	3	1.5	1.80278			
6	4	2	2.23607			

Table 3.

Copyright © 2012 Pearson Education, Inc. Publishing as Addison-Wesley

napter 10 Differential Equations

Euler's Method

Example 1 of this section of the text can be completed using Excel. Set up the sheet with columns for the independent variable *x*, the approximate solution variable *y*, the actual solution *f(x)*, and the error *y-f(x)*. You also need a single cell containing the step *h*. Table 1 shows the set-up with the proper initial values in the range `A1:E4`.

x	y	f(x)	y-f(x)	h
0.0	1.5	1.5	0.0	0.1

Table 1.

The contents of all the cells in the second row are numbers except for `D2`, which contains the difference formula `=B2-C2`. The third row will be all formulas, as seen in the following:

Cell	Contents
A3	`=A2+$E$2`
B3	`=B2+$E$2*(A2-2*A2*B2)`
C3	`=0.5+Exp(-(A2^2))`
D3	`=B3-C3`

Notice the absolute references for cell `E2`, as you want to use the same step size *h* throughout the calculations. Now select the range `A3:D3` and fill down nine rows. Your results should be identical to Table 2.

x	y	f(x)	y-f(x)	h
0	1.5	1.5	0	0.1
0.1	1.5	1.49005	0.00995	
0.2	1.48	1.460789	0.019211	
0.3	1.4408	1.413931	0.026869	
0.4	1.384352	1.352144	0.032208	
0.5	1.313604	1.278801	0.034803	
0.6	1.232243	1.197676	0.034567	
0.7	1.144374	1.112626	0.031748	
0.8	1.054162	1.027292	0.026869	
0.9	0.965496	0.944858	0.020638	
1	0.881707	0.867879	0.013827	

Table 2.

Copyright © 2012 Pearson Education, Inc. Publishing as Addison-Wesley